AN INTRODUCTORY GUIDE TO
FLOW MEASUREMENT

Other titles in this series:

An Introductory Guide to Flow Measurement

(2nd Edition)

ROGER C. BAKER

Visiting Professor
Cranfield University

Visiting Industrial Fellow
Institute for Manufacturing
Cambridge University

**Professional
Engineering
Publishing**

Professional Engineering Publishing Limited
London and Bury St Edmunds, UK

First published 1989
Second edition published 2002

ISBN 1 86058 348 2

A CIP catalogue record for this book is available from the British Library.

Printed by The Cromwell Press, Trowbridge, Wiltshire, UK

Series Editor's Foreword

As an engineer I often feel the need for an introductory guide to some aspect of engineering outside my own area of knowledge. Professional Engineering Publishing welcomed the idea of publishing a series of such guides to follow on from the first edition of this one. We hope that the series will continue to provide engineers with an easily accessible set of books on common and not-so-common areas of engineering. Each author brings a different style to his subject, but some features of the original volume, such as conciseness and the emphasis of certain sections by shading, have been retained.

The series is designed to be suitable for practising engineers, for degree course students, for instrument technicians, for design engineers and those responsible for specifying plant, and for teachers and researchers. In each case the books give a starting point and a clear explanation of the subject to allow the reader to assess commercial literature, to follow up more advanced technical books, and to have more confidence in dealing with those who claim an expertise in the subject.

I am obviously delighted that the series' popularity has resulted in this second edition of my original volume in the series.

We hope that the series will continue to find a welcome with engineers and we shall value reactions and any suggestions for future volumes in the series.

Roger C. Baker
St Albans
December 2001

Disclaimer

Every effort has been made in preparing this book to provide accurate and up-to-date data and information that is in accord with accepted standards and practice at the time of publication and has been included in good faith. Nevertheless, the author, editors, and publisher can make no warranties that the data and information contained herein are totally free from error, not least because industrial design and performance are constantly changing through research, development, and regulation. Data, discussion, and conclusions developed by the author are for information only and are not intended for use without independent substantiating investigation on the part of the potential users. The author, editors, and publisher therefore disclaim all liability or responsibility for direct or consequential damages resulting from the use of data, designs, or constructions based on any of the information supplied or materials described in this book. Readers are strongly advised to pay careful attention to information provided by the manufacturer on any equipment that they plan to use and should refer to the most recent standards documents relating to their application. The author, editors, and publisher wish to point out that the inclusion or omission of any particular device, design, application, or other material in no way implies anything about its performance with respect to other devices, etc.

Contents

Author's Preface

This book is aimed at the busy practising engineer, who is faced with a flow measurement problem and requires enough information to enable him or her to assess the advice received from manufacturers, and to contribute to discussions with experts.

The first edition of this book has been widely used over its 13 years in print, and appeared to fulfil the aim. In producing a second edition, I have attempted to retain the succinctness of the first edition. This has been possible for two reasons: some material which was of marginal value has been removed, and for the reader who needs more information, this is now readily available in two other books (*Flow Measurement Handbook*, 2000; *An Introductory Guide to Industrial Flow*, 1996). In this book, therefore, I have attempted to select from the wide ranging subject the information which will allow the reader to understand the essential background technology, to obtain the essentials of the flowmeter operating principles, to have some guide as to the likely performance of a particular instrument, and to weigh up the advantages and disadvantages.

This book is about existing flowmeters, their operation, installation, and application advantages and disadvantages. As such it is concerned with the mechanical and fluid mechanical aspects. The output signal, if electrical, is likely to be 4–20 mA, a frequency or pulse signal, or one of the digital protocols. Most manufacturers will offer a standard range of output signals, although these may be selectable options. I have not attempted to provide such information, since it is changing continually, and should not cause problems for most users.

The style of the book has been developed since the first edition. Shading is again used to highlight the key points, but in this edition, the figures and tables are given their own titles and should stand out sufficiently on their own. The book stemmed originally from my lectures at Cranfield, and it has been used intensively for the same purpose over its 13 years' life. The result is that some of the lessons I have learned from lecturing from the book have found their way into this edition.

Chapters 1 and 2 set the background in terms of accuracy, flow behaviour, fluid parameters, and calibration techniques. Chapter 3 attempts to guide the reader to the selection of the most appropriate meter for his or her application. The main considerations are listed, and the flowmeter types are tabulated with operating ranges and likely performances. From this chapter the reader may wish to turn to the sections of most immediate interest. Chapters 4 to 7 review as many designs and techniques as possible within the scope of this short book and Chapter 8 looks at recent developments and likely future trends.

I have attempted to provide a fair and correct description and assessment of each flowmeter type and its limitations. However, it should be recognized that some instruments may out-perform the guidelines and others may fail to meet them. The reader should, therefore, accept the values given with caution and seek further advice when necessary.

The dominant reflection, having completed this book, is that almost every statement needs much more qualification than it is possible to give, and every attempt to generalize is dangerous. Therefore, reader, do not proceed uncritically, and if you find this book useful, but can suggest corrections and improvements, the author will be pleased to hear from you.

Roger C. Baker
St Albans
December 2001

Acknowledgements

In writing this book I have been particularly conscious of the debt which I owe to many colleagues, in academia and industry, and to my students, from whom I have learned. I have also been privileged to listen to many colleagues lecturing on flow measurement, and to benefit from their writings.

I am particularly grateful to colleagues who read and commented on sections of the manuscript: Wes Allen (Endress+Hauser), Alan Best and Trevor Billington (Elster Jeavons), Chris Gimson (Endress+Hauser), Andy Nurse (Quadratics Consulting), Michael Reader-Harris (NEL), John Salusbury (Endress+Hauser), Ben Weager (Danfoss), and Yousif Hussain (Krohne). I would, especially, like to thank Gary Oddie (Schlumberger) who read and commented on all but two of the chapters. I am extremely grateful to each one of them for taking time to do this, and for the constructive comments and insights they offered. Of course, I bear full responsibility for the final script.

I have been pleased to retain close connections with Cranfield, as a Visiting Professor in the department that has developed from the one I originally set up in 1977. I am grateful to my successor, Professor Michael Sanderson, and to the University for the continuing collaborative links there. I have now returned to Cambridge University Engineering Department and am grateful to Professor Mike Gregory and to the Department for the opportunity this affords me to continue my research, to develop my industrial links, and to explore the manufacturing implications of producing high-quality instruments.

Cambridge University Press kindly gave permission for the use of Figures 1.7, 4.2, 5.18, 5.20, 5.23, 6.1, 6.5, 7.2, and Table 3.1 which are taken from Baker (2000).

I am grateful to Professional Engineering Publishing and particularly to Sheril Leich for encouraging me to produce a second edition of this Introductory Guide. However, without the unwavering encouragement and support of Liz, not to mention her meticulous advice on the manuscripts, I doubt if I should ever have succeeded.

to

Liz,

Rachel, John and Mark,

Sarah and Paul,

Rebekah and Isaac

Nomenclature

A	Area of duct; as defined in relation to equation (4.3); area of target; constant
A_*	Throat area for critical (sonic) nozzle
a	Constant
a_ε	Constant in equation (4.4)
B	Bias; magnetic flux density
b	Constant
b_ε	Constant in equation (4.4)
C	Coefficient of discharge
C_0, C_1, C_2	Tracer concentrations
C_*	Critical flow function for critical (sonic) nozzle
C_{*i}	Critical flow function for perfect gas
C_D	Drag coefficient for target meter
c	Sound speed
c_i	Sensitivity coefficient for input quantity x_i
c_p	Specific heat
D	Pipe diameter
d	Orifice or throat diameter; bluff body dimension
E	$\sqrt{[1/(1 - \beta^4)]}$
F	Force on target meter
f	Frequency
f_t	Transmitted frequency
Δf	Frequency difference
g	Gravitational acceleration
h, H	Height/head
K	Loss coefficient; pulses/unit volume; constant
k	Coverage factor to obtain uncertainty; constant
L	Aerofoil lift; bend spacing; length of leakage gap in positive displacement (PD) meter; ultrasonic beam/path length
L_1	l_1/D
L_2'	l_2'/D
l	Length of pipe; length of PD meter rotor; axial beam spacing
l_1	Distance of the upstream tapping from the upstream face of the orifice plate
l_2'	Distance of the downstream tapping from the downstream face of the orifice plate
$\mathbf{M}$	Molecular weight

M_2'	As defined for equation (4.3): the distance between the downstream tapping and the downstream face of the orifice plate divided by the dam height
m	Area ratio of throat to pipe
N	Number of blades on turbine meter wheel
n	Number of readings; as used in equations (1.11) and (4.9)
p	Pressure
p_o	Upstream stagnation pressure for critical nozzle
Δp	Pressure difference
Δp_{AB}	Pressure difference between points A and B, and similarly for points AC, AD, and BD
Q_h	Heat transfer
$\bar{q}$	Mean of a series of repeat readings
q_h	Heat supplied
q_j	Repeat readings
q_m	Mass flowrate
q_v	Volumetric flowrate
q_{v1}	Volumetric flowrate of injected tracer
q_{vp}	Volumetric flowrate through pump in hydraulic Wheatstone method
$\mathbf{R}$	Universal gas constant
R	Pipe radius; wheel radius
Re	Reynolds number
r	Radius
r_h	Hub radius
r_t	Tip radius
δr	Small radial movement of mass δm
$\delta r'$	Small element of vibrating tube
S	Experimental standard deviation; sensitivity of electromagnetic flowmeter
St	Strouhal number
s	Standard deviation; spring constant
T	Temperature; torque
T_o	Upstream stagnation temperature for critical nozzle
ΔT	Temperature difference
t	Leakage gap clearance; time
Δt	Timed difference of ultrasonic up- and downstream pulses
U	Expanded uncertainty
ΔU	Potential difference between electrodes
$u(x)$	Uncertainty for a measurement x
$u_c(y)$	Combined uncertainty of y, the output quantity
V	Voltage; volume of a pipe between sampling points

V_1	Velocity at station 1
V_{ax}	Axial flow velocity
V_b	Blade velocity
V_o	Centre-line velocity
V_{rel}	Velocity, relative to the moving blade, of fluid passing over a turbine blade
$\overline{V}$	Mean velocity
v	Specific volume; volume of injected water with tracer
w	Width of bluff body
X	Axial spacing between ultrasonic transducers; angular momentum
x_i	Measurement value
$\bar{x}$	Mean value of measurements
y	Output quantity
z	Height above datum

Greek symbols

β	d/D; blade angle
γ	Ratio of specific heats
δ	Calibration correction; region lost by PD meter vanes
ε_1	Expansibility factor based on pressure at station 1
θ	Beam/path angle
κ	Isentropic exponent
μ	Dynamic viscosity
ν	Kinematic viscosity
ρ	Fluid density
ρ_1	Fluid density at station 1 (upstream pressure tapping) for differential pressure meters
τ	Time period
ω	Rotational speed or frequency in radians per second

Subscripts

1, 2, 3	Position in meter, for instance of pressure tapping
ax	Axial, for instance the velocity along the direction of the pipe in a turbine meter
b	Blade
i	Indicating input
i, j	To indicate one of a number of values
rel	Relative to the blade, as in velocity in the turbine meter
t	Transmitted, as in Doppler flowmeter transmission frequency

CHAPTER 1

Introduction

1.1 INTRODUCTION

Do you need a flowmeter?

Momentum, volumetric, or mass flow measurement

- which do you need?

How accurate?

- accuracy costs money.

Maintenance

- can you afford it?

Specification

- have you provided the manufacturer with the complete specification for your needs?
- has the manufacturer given you all the information you want?

Before you install a flowmeter, ask why you are doing so. Do you *really* need a flowmeter? Some years ago I was working on a flowmeter design to monitor flows in a very sensitive installation. Because flowmeter failure was unacceptable and two flowmeters could contradict, three flowmeters were proposed for each flowmeter position, and then four. At this stage it was considered that the monitoring might be done better with thermometers rather than flowmeters! More recently, the move of the UK water industry from public to private control has highlighted the danger of installed flow-meters which have not been used for some time and for which no records have been kept. Their existence may be more of a problem than their lack would ever have been!

Do you need to measure the volumetric flow of the fluid ('fluid' in this book will be used to mean either liquid or gas, or combinations of solids, liquids, and gases) or do you require the mass flow? In this book flow-meters are subdivided into momentum-, volume- and mass-sensing cate-gories. The flowmeters which respond to momentum are density dependent and a knowledge of the density will be required to obtain either volume or mass flow.

If you do indeed need a flowmeter, how accurate does it need to be? Accuracy costs money, not only in the initial cost of the instrument and an adequate installation, but also in the maintenance of the required accuracy. If the installation will not allow adequate maintenance, or if the flowmeter maintenance is likely to be the first casualty of financial economies, it is important to question the advisability of installing the instrument in the first place. It may also be worth considering whether another instrument, measuring a quite different parameter, could be used instead to give an indication of plant behaviour. Indeed some of the most recent work on measuring flow sees the use of clever computational methods to measure flow without a flowmeter! Virtually all flowmeters are available with standard electrical output, but the development of smart and then intelligent instruments, and the arrival of various fieldbus specifications all have an effect on the flowmeter output. A few types still have a predominantly mechanical read-out, but it seems likely that the major advances in signal processing and interpretation will change the whole nature of the electrical end of flowmeters in the next few years.

Adequate communication between the manufacturer and the purchaser of flowmeters is essential in ensuring that the meter is appropriate to the application. Although there are some manufacturers with one or two specialist meters, the majority now appear to offer a range for most eventualities. It is important to make sure that the application specification is known to the manufacturer, and that the manufacturer in turn makes it clear in what areas the meter does not meet that specification.

In this book an introduction is given to meters which measure the flow of fluids in pipes. For flows in open channels the reader is referred to Herschy (1995). For other areas outside the scope of this book, the reader is referred to Baker (2000) where additional references may be found.

In the next few paragraphs we look first at the meaning of accuracy and the key words used to describe it, and then at fluid mechanical, and other factors which will affect the accuracy of the installed flowmeter.

It is useful to be able to change from one unit of flowrate to another, and Table 1.1 provides some conversions. In the table, there is also an indication of the velocity of flow in a pipe for a given flowrate and pipe diameter. This can be useful in giving a feel for the likely maximum and minimum flowrates which a meter can sense. Typically, for liquids flowrates are in the range 0.1–10 m/s but for gases may be 30 m/s or higher.

Note that in this book we use the international metric system of units (SI).

Table 1.1 Flowrates and velocities for various pipe sizes

Mean velocity (m/s) in a circular pipe of diameter: (columns 10 mm – 2000 mm)

	m^3/h*	l/min	Imperial gal/min	US gal/min	ft^3/min	10 mm	25 mm	50 mm	100 mm	200 mm	500 mm	1000 mm	2000 mm
Very low	10^{-3}	0.017	3.7×10^{-3}	4.4×10^{-3}	5.9×10^{-4}	3.5×10^{-3}	5.7×10^{-4}	1.4×10^{-4}	3.5×10^{-5}				
	10^{-2}	0.17	3.7×10^{-2}	4.4×10^{-2}	5.9×10^{-3}	3.5×10^{-2}	5.7×10^{-3}	1.4×10^{-3}	3.5×10^{-4}	8.8×10^{-5}	1.4×10^{-5}		
	0.1	1.7	0.37	0.44	5.9×10^{-2}	0.35	5.7×10^{-2}	1.4×10^{-2}	3.5×10^{-3}	8.8×10^{-4}	1.4×10^{-4}	3.5×10^{-5}	
	1	17	3.7	4.4	0.59	3.5	0.57	0.14	3.5×10^{-2}	8.8×10^{-3}	1.4×10^{-3}	3.5×10^{-4}	8.8×10^{-5}
	10	170	37	44	5.9	35	5.7	1.4	0.35	8.8×10^{-2}	1.4×10^{-2}	3.5×10^{-3}	8.8×10^{-4}
	100	1700	370	440	59	350	57	14	3.5	0.88	0.14	3.5×10^{-2}	8.8×10^{-3}
	1000	1.7×10^{4}	3700	4400	590		570	140	35	8.8	1.4	0.35	8.8×10^{-2}
	10^{4}	1.7×10^{5}	3.7×10^{4}	4.4×10^{4}	5900				350	88	14	3.5	0.88
	10^{5}	1.7×10^{6}	3.7×10^{5}	4.4×10^{5}	5.9×10^{4}					880	140	35	8.8
Very high	10^{6}	1.7×10^{7}	3.7×10^{6}	4.4×10^{6}	5.9×10^{5}							350	88

*Since water has a density of 1000 kg/m^3 (approximately), the mass flowrate (in kg/h) of water may be obtained by multiplying this column by 1000 (values to two significant figures).

1.2   ACCURACY

Accuracy
- the truthfulness of an instrument.

Repeatability
- the closeness of the agreement between measurements from the same instrument.

Uncertainty
- the quality of the measurement;
- in this book uncertainty will be given as a percentage of actual flowrate.

Confidence level
- the probability of the reading falling within certain limits;
- a 95 per cent confidence level is common in flow measurement.

Linearity
- the closeness to a linear response.

Range
- the maximum and minimum flowrates for which the uncertainty, linearity, etc. are applicable.

Turndown ratio
- the ratio of maximum to minimum flowrate range values.

Systematic and random errors

Evaluation of standard uncertainty
- the *Guide* (BIPM *et al.* 1993) and UKAS (1997) cf. Baker (2000).

There is much confusion about the various terms used to indicate the performance and quality of flowmeters. The terms are discussed below.

Accuracy
It is generally accepted that accuracy refers to the truthfulness of the instrument. An instrument of high accuracy more nearly gives a true reading than an instrument of low accuracy. Accuracy, then, is a quality of the instrument. It is common to refer to a measurement as accurate or not, and we understand what is meant. However, it is unsatisfactory to refer to a measurement's accuracy of, say, 1 per cent, when presumably we mean that the instrument's reading will lie within a band of 99–101 per cent of the true reading. Thus, accuracy refers to the ability of an instrument to give a reading close to the absolute value (traceable to a national metrology standard).

Repeatability
The performance of a meter in a process plant, or other control loop, may not require the absolute performance level that is needed when buying and

selling the liquid or gas, but may require repeatability within bounds defined by the process. A flowmeter with good repeatability is one for which the probable difference between two readings is small. Although not favoured for flowmeters, in some instrumentation the words *repeatability* and *reproducibility* are both used. The former is used where we are interested in the difference between successive readings under constant conditions with the same observer and a small elapsed time, whereas the latter implies identifiable change of conditions such as a different observer, location, or long elapsed time. It will be important, when we use the word *repeatability* in flowmetering, to be clear with which meaning we are concerned.

Precision
This is the qualitative expression for repeatability, should not take a value, and should not be used as a synonym for accuracy.

Uncertainty
This word is properly used to refer to the quality of the measurement, and we can correctly refer to an instrument reading having an uncertainty of ±1 per cent, provided we also define under what circumstances this is valid.

Confidence level
It is not satisfactory to state an uncertainty without also indicating the probability. This is given as a confidence level and, usually, for flow measurement, this is 95 per cent. We shall assume this level in this book. Simply put, this means that out of 20 readings we would expect, on average, that 19 would fall within the uncertainty band ($19/20 = 95/100 = 95$ per cent).

Linearity
This term is of use for instruments that give a reading approximately proportional to the true flowrate over their specified range. It then refers to the closeness within which the meter achieves a truly linear or proportional response. It is usually defined by stating the limits, for example, ±1 per cent of flowrate, within which the response lies over a stated range. With modern signal processing this is probably less important. It is a special case of *conformity* to a curve.

Range and turndown ratio
An instrument should have a specified range, or turndown, over which its performance can be trusted. Without such a statement the values of uncertainty, linearity, etc. are inadequate. It is important to note whether the values of uncertainty, linearity, etc. are related to the actual flowrate or to the full scale deflection (FSD) (sometimes referred to as URV – upper range value).

For instance, ±1 per cent uncertainty on rate for an instrument with a 20:1 turndown would be impressive, but ±1 per cent of full scale for a 100:1 turndown might raise questions about the performance at the bottom end of the range.

Systematic and random error

Random error, the random part of the experimental error, causes scatter, and reflects the quality of the instrument design. It is the part that cannot be calibrated out, and the smaller it is the more precise the instrument.

Systematic error, according to flowmeter usage, is that which is unchanging within the period of a short test with constant conditions. However, there may be longer term drift, or change, with changes in conditions, which results in a change in the *bias*, another term used for systematic error. The systematic error may result from the meter being calibrated, the calibration equipment, the reference source, the operational procedure and environmental factors, and it may be stable or subject to drift.

Evaluation of standard uncertainty

The reader is referred to the *Guide to the Expression of Uncertainty in Measurement* (BIPM *et al.* 1993) and to UKAS (1997) which set out in detail the current philosophy for estimating standard uncertainty. A brief introduction, but fuller than here, can be found in Baker (2000). Essentially the overall uncertainty is achieved by starting with the standard uncertainty for a measurement x, $u(x)$, in the units of the measurement (sometimes referred to as δx or Δx).

For a series of repeat readings q_j, the *experimental standard deviation* is given by

$$s(q_j) = \sqrt{\frac{1}{(n-1)} \sum_{j=1}^{n} (q_j - \bar{q})^2} \qquad (1.1)$$

The estimated standard deviation of the uncorrected mean is

$$s(\bar{q}) = \frac{s(q_j)}{\sqrt{n}} \qquad (1.2)$$

If it is not convenient to repeat the measurement many times during a calibration it may be possible to use an earlier value. The *standard uncertainty* is then given by

$$u(x_i) = s(\bar{q}) \qquad (1.3)$$

We can combine random uncertainties conservatively by arithmetic addition, but UKAS (1997) following the *Guide* takes the square root of the sum of the squares. Thus the *combined standard uncertainty* is obtained as

$$u_c(y) = \sqrt{\sum[c_i u(x_i)]^2} \qquad (1.4)$$

where y is the output quantity, c_i is a sensitivity coefficient (derived from the partial derivative – see UKAS 1997 for details) for each input quantity which ensures consistent units, and x_i are the input quantities.

The *Guide* (BIPM *et al.* 1993) replaces the use of Student's t with a coverage factor, k, to obtain the expanded uncertainty

$$U = ku_c(y) \qquad (1.5)$$

and recommends a value of $k = 2$ which gives a confidence level of 95.45 per cent taken as 95 per cent assuming a normal distribution of data.

1.3   FLOWMETER SYSTEMS

Definition of an ideal flowmeter
- a group of linked components that will deliver a signal uniquely related to the flowrate or quantity of fluid flowing in a conduit, despite the influence of installation and the operating environment.

Operating envelope
- related to full scale or rate?
- uncertainty and confidence level.

Meter factors
- *K* factor.
- Meter factor.

The object of installing a flowmeter system is to obtain a measure of the flowrate, usually in the form of an electrical signal, which is unambiguous and within specified uncertainty limits. This measurement should be negligibly affected by the inlet and outlet pipework systems and the operating environment.

For fixed conditions in which only the flowrate is varied the ratio of measured to true flowrate should be a function of flowrate only. Ideally it will be unity (that is, linear); in practice it will lie within the sort of envelope shown in Fig. 1.1. This indicates a flowmeter performance for which the uncertainty is within 1 per cent for a 10:1 turndown ratio. This envelope might be a consequence of the instrument being calibrated at a point near the top of the range and another at 50 per cent. The manufacturer, who probably carried out the calibration, knew from production control and sample testing that with a single adjustment the performance would lie within this envelope. It is apparent that this could equally be seen as a linearity envelope. Such performance in a flowmeter would be quite good. If, on the other hand, the uncertainty was defined in terms of full-scale flowrate or upper range value (URV), then the envelope would become trumpet shaped, with an increasing uncertainty as a percentage of the actual flowrate, as flow is reduced (Baker 2000). A 1 per cent uncertainty based on full-scale flowrate would give a 10 per cent uncertainty at 10 per cent of the flowrate, if based on the actual flowrate. A further variant is that some flowmeters are defined in terms of a percentage of actual flowrate, but this changes from a lower value for the top of the range to a higher value for low flowrates.

It is very important, therefore, that the actual operating envelope of the flowmeter is clear to the user and that it meets the user's specification. Within this operating envelope there will be a flowmeter characteristic, which will be obtained by calibration over the operating range, and which could have a considerably lower value of uncertainty than the manufacturer's published figures.

It is useful to introduce two factors which define the response of flowmeters. They are most commonly used for linear flowmeters with pulse output. The K factor is the number of pulses per unit quantity. In this book we shall take it as the number of pulses per unit volume when dealing with turbine and vortex meters:

$$K = \frac{\text{pulses}}{\text{true volume}} \qquad (1.6)$$

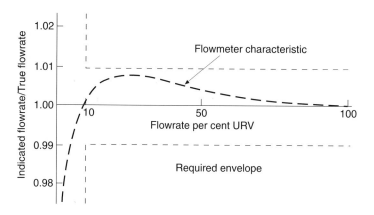

Fig. 1.1 Flowmeter performance envelope

while the meter factor is usually defined as

$$\text{Meter factor} = \frac{\text{true volume}}{\text{indicated volume}} \tag{1.7}$$

The reader should keep a wary eye for other definitions of meter factor such as the reciprocal of the K factor.

1.4   FLOW IN PIPES

One almost invariable installation effect is caused by the upstream and downstream pipework. We, therefore, consider first the nature of the undisturbed flow in a pipe, and then the effect of various pipe fittings.

Flow similarity
- Reynolds experiment.
- Laminar and turbulent flow profiles.

$$\text{Reynolds number} = \frac{\rho V D}{\mu} = \frac{\text{inertial forces}}{\text{viscous forces}}$$

- Pipe wall roughness.
- Shape of flow profiles.
- Effect of upstream pipework on flow profiles.

One of the most important concepts in fluid mechanics is similarity of flows. Much of what we shall discuss in this book is concerned with the concept that flows in two different pipes are similar in various ways providing certain relationships are satisfied.

Reynolds experiment – laminar and turbulent flow profiles
Osborne Reynolds (1883) set up an experiment in which he demonstrated the laminar/turbulence transition (Fig. 1.2). In this experiment, if a visualization technique such as a dye streak is produced along the axis of the pipe through which water flows, flow changes may be observed. If the flow into the pipe is smooth, then for low flowrates the dye streak shows very little change. With great care over the stillness of the entering flow, and lack of vibration in the rig, the undisturbed nature of the flow can be sustained up to quite high flows. However, without such careful precautions, it is found that there is a critical condition below which the streak will always be well defined, but above which any disturbance in the flow will be magnified and the dye streak will be broken up by eddies. The condition is given by the Reynolds number,

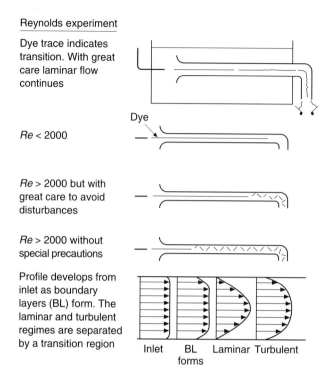

Reynolds experiment

Dye trace indicates
transition. With great
care laminar flow
continues

Dye

$Re < 2000$

$Re > 2000$ but with
great care to avoid
disturbances

$Re > 2000$ without
special precautions

Profile develops from
inlet as boundary
layers (BL) form. The
laminar and turbulent
regimes are separated
by a transition region

Inlet   BL   Laminar Turbulent
         forms

Fig. 1.2 Indication of the main features of Reynolds' experiment

Re, which provides an indication of the ratio of inertia forces in the flow to viscous forces and is given by

$$Re = \frac{\rho V D}{\mu} \tag{1.8}$$

where ρ is the fluid density, μ the viscosity, V the velocity, and D the pipe diameter in consistent units. For *Re* less than about 2000 the flow in the pipe is laminar and all the fluid travels in a direction parallel to the pipe axis. Above this value of *Re* small disturbances will normally grow forming turbulent eddies so that, superimposed on the axial flow, there are circulating eddies of many sizes with velocities up to about one-tenth of the axial velocity for a smooth pipe. The effect of these turbulent eddies is to mix the flow and to create a turbulent profile in the pipe which is more uniform. Common fluids such as air and water are termed Newtonian fluids, since μ is constant for constant temperature and pressure and independent of shear stress in the fluid. For non-Newtonian fluids the viscosity is a function of the shear caused by the flow.

Flow similarity – Reynolds number
The importance of flow similarity is that, for two geometrically similar pipes, the flow behaviour will be the same for equal values of *Re* in each pipe. One aspect of similarity is that the wall roughness is similar. The average height of roughnesses is usually given as a fraction of the pipe diameter, and for the pipes to be similar, this roughness must also be in proportion to the pipe sizes. If we specify, therefore, for calibration of a flowmeter, a certain value of *Re* and similar pipes, then we know that the profile will be well defined.

It turns out that the roughness, if small enough, will be within a very thin layer of flow next to the pipe wall, and will not affect the profile of the flow. If it is larger than this value, then it will have an effect, and if large enough will dominate the profile shape.

Flow profile shapes
We find that for laminar flow the pressure drop in a pipe is proportional to the velocity, and the profile shape is parabolic given by

$$V = V_0[1 - (r/R)^2] \tag{1.9}$$

and

$$\overline{V} = V_0/2 \tag{1.10}$$

where V is the velocity in the pipe at radius r, R is the pipe radius, $\overline{V}$ is the mean velocity in the pipe, and V_0 is the velocity on the pipe axis (for the derivation see Baker 1996).

For turbulent flows the profile can be approximated by

$$V = V_0(1 - r/R)^{1/n} \tag{1.11}$$

and

$$\overline{V} = \frac{2n^2V_0}{(1+2n)(1+n)} \tag{1.12}$$

and the value of n can be related to the Reynolds number according to Table 1.2.

Between the turbulent onset and the laminar flow there is a transition when the flow alternates randomly between laminar flow and patches of turbulent flow. The last two lines in Table 1.2 indicate that a feature of these profiles is that the ratio of the velocity at about the three-quarter

Table 1.2   **Experimental values of *n* for various *Re*, and the ratio of local to mean velocity at various radii**

Re	4×10^3	2.3×10^4	1.1×10^5	1.1×10^6	2.0×10^6 to 3.2×10^6
n	6.0	6.6	7.0	8.8	10
$V/\overline{V}$ (at $r = 0.75R$)	1.003	1.004	1.004	1.005	1.005
$V/\overline{V}$ (at $r = 0.758R$)	0.998	0.999	1.000	1.002	1.002

radius point to the mean velocity is approximately unity. In fact if the radial position is taken as about 0.758, then the ratio is even closer to unity. This leads to the possibility of using a local velocity probe, set at this point in a fully developed turbulent profile, to deduce the mean velocity.

Figure 1.3 shows approximate profile shapes obtained from equations (1.9) and (1.11). These profiles all assume that there is a sufficient length of straight pipe to ensure that they are fully developed. For turbulent flow this is usually taken to be at least 60 diameters (60*D*). It should be noted that the fluid next to the wall of the pipe has zero velocity. This is known as the non-slip condition, and means that the velocity in Fig. 1.3 goes to zero at the top and bottom of the profile. It is also worth noting that turbulent flow will exist in most industrial flowmeter applications. For instance, water at 2 m/s in a 25 mm pipe has a Reynolds number of 50 000, and air at 20 °C for the same velocity and pipe diameter has a Reynolds number of 3250.

Some fluids may display non-Newtonian behaviour, by which is meant that the viscosity, μ, is not constant for a given temperature and pressure, but is a function of the shearing conditions in the fluid and/or the rate of change of shear with time. For example, the shear caused by whipping cream causes a change in its consistency. Some fluids exhibit a critical stress before flow takes place and others exhibit time-dependent effects. This may influence the flowmeter performance by its effect on flow profile shape within a pipe, and

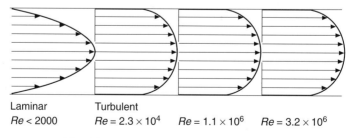

Laminar   Turbulent

$Re < 2000$ $Re = 2.3 \times 10^4$ $Re = 1.1 \times 10^6$ $Re = 3.2 \times 10^6$

Fig. 1.3 Laminar and turbulent pipe profiles

also by its effect on the local flow patterns around the internal geometry of the flowmeter. It is unlikely that any data will be available on the effects of a non-Newtonian fluid on a particular flowmeter unless the manufacturer has carried out tests.

Effects of upstream pipework on flow profiles
So far we have considered ideal flow profiles resulting from the laminar or turbulent flow in a long enough piece of pipe to result in a fully developed profile. In most flowmeter installations it will be very difficult to ensure that the flow profile is fully developed. More often the flow profile will be distorted by a bend upstream or an area change due to a contraction, expansion, or valve in the pipework. This profile distortion will affect the flow for several diameters downstream and such fittings will also lead to additional flow losses. We have not discussed the question of flow losses in this book, but the reader is referred to Miller (1978) for an excellent source of information on these.

Bends. Due in part to its importance industrially, the flow in a bend has been quite extensively documented and the effects on flowmeters have been measured. Figure 1.4a shows in diagrammatic form the effect of a bend on the flow profile. The higher velocity fluid at the centre of the pipe moves to the outside of the bend while the lower velocity fluid at the wall moves to the inside of the bend. The result at the outlet of the bend is shown by a region of lower velocity and a region of higher velocity. This is clearly rather different from the turbulent profile with its axisymmetry in Fig. 1.3. This profile will, in most cases, affect the registration of a flowmeter which has been calibrated on a fully developed flow profile.

Two bends in succession will have a greater effect than one bend in most cases, but the worst situation is when the bends are in perpendicular planes. The result is that the flow develops swirl, the rotation of the flow in the pipe, as shown in Fig. 1.4b. This happens because the low velocity fluid is deposited on one side of the second bend and the high velocity fluid on the other side, and when these have passed through the bend they have a component of rotation in the outlet pipe. The problem with this is that swirl tends to be very long lasting and most flowmeters are seriously affected by its presence.

Changes in area. Flow through an expansion may lead to a submerged jet flow (Fig. 1.5). As it issues from the smaller inlet pipe, the flow sheds vorticity which creates the recirculation zone. The profile in the larger outlet pipe may be peaked due to the submerged jet.

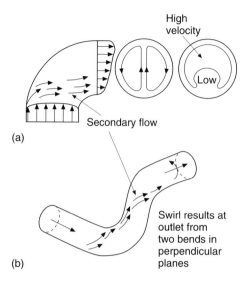

Fig. 1.4 Flow in bends: (a) single bend; (b) two bends in perpendicular planes

The flow in a contraction, although experiencing a decreasing pressure gradient and resulting in a flattened profile, may create separation zones as the flow emerges from the contraction into a succeeding parallel section.

If the pipe is partially blocked, by a thermometer pocket (thermowell) for example, this could disturb the flow and could also lead to vortex shedding, the alternate flow around the pocket on each side, rather as we shall find in the vortex shedding flowmeter.

Valves. There is a useful rule which is never to position a flowmeter (within, say, 60D) downstream of a valve. The problem with a valve upstream of a flowmeter is not only that it will cause a reduction in the accuracy of the flowmeter but also, if the valve opening is varied, that it will not even be a constant reduction.

1.5   EFFECT ON FLOWMETERS

We shall find that, for many flowmeter types, data are available that indicate the likely size of errors resulting from the presence of pipe fittings upstream of the flowmeter, and the distance in diameters necessary to keep the errors below certain values. The most documented example of this is for the orifice plate flowmeter. ISO 5167 provides a useful table for various fittings upstream of an orifice plate of various diameter ratios. In Table 3.2 the

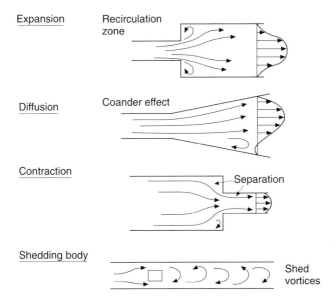

Fig. 1.5 Flow with area change

installation distances have been adapted into a simpler table to allow use with other flowmeters.

Flow straighteners and conditioners
Because of profile distortion resulting from upstream fittings, attempts have been made to produce a pipe fitting which will return the flow from its distorted form to the fully developed datum form.

The distinction that should be made between flow straighteners and conditioners, is that a straightener is there to reduce swirl, while a conditioner is there to correct the shape of the profile. Figure 1.6 shows a selection of flow straighteners. It is essential that straighteners are made with great care otherwise the device can introduce swirl rather than removing it. That in Fig. 1.6a uses a star of flat plates, that in Fig. 1.6b a bundle of straight small tubes within the main pipe, and in the combined Zanker flow conditioner and straightener (Fig. 1.6d) a honeycomb structure is used. The conditioners have moved on from the design of Zanker, and tend now to be based on the perforated plate (Fig. 1.6c). They consist of a thick plate (Figs 1.6e and f), possibly with straightening tabs and settling regions. An alternative approach is to create strong vorticity to enable the profile to settle down more quickly (see Baker 2000 for references).

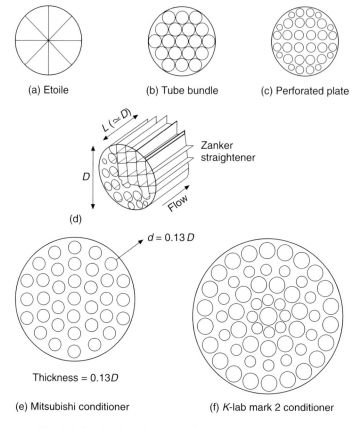

Fig. 1.6 A selection of flow straighteners and conditioners

The devices that seem to be most promising are the perforated-plate-type conditioners.

1.6 ESSENTIAL EQUATIONS OF FLOW

The derivation of the following equations can be found in textbooks of fluid mechanics, and they are given in a little more detail in Baker (1996). Here they are quoted without further justification.

Continuity equation
This states that the same mass flows through two cross-sections of a pipe which is unbranched and continuous (Fig. 1.7):

$$q_m = \rho_1 A_1 V_1 = \rho_2 A_2 V_2 \tag{1.13}$$

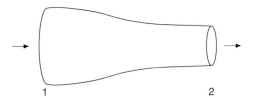

Fig. 1.7 Varying area duct to show stations referred to in equations (1.13) to (1.16) (from Baker 2000 reproduced by kind permission of Cambridge University Press)

where q_m is the mass flowrate, ρ is the density, A is the duct area and V is the velocity. Subscripts 1 and 2 indicate that the values are taken at the first and second stations respectively.

For incompressible flow with constant density this becomes

$$q_v = A_1 V_1 = A_2 V_2 \qquad (1.14)$$

where q_v is the volumetric flowrate.

Bernoulli's equation
For an incompressible fluid Bernoulli's equation is

$$\frac{V_1^2}{2} + \frac{p_1}{\rho} + gz_1 = \frac{V_2^2}{2} + \frac{p_2}{\rho} + gz_2 \qquad (1.15)$$

where p is the pressure, g is the gravitational constant, and z is the height above datum.

Pressure change equation
We can use equations (1.14) and (1.15) to obtain the form of the equation for ideal incompressible volumetric flow through a contracting duct such as that in Fig. 1.7 where $z_1 = z_2$:

$$q_v = E(\pi / 4)d^2 \sqrt{(2\Delta p\rho_1)} \qquad (1.16)$$

where

$$E = 1 / \sqrt{(1 - \beta^4)} = 1 / \sqrt{(1 - m^2)}$$
$$\beta = d/D$$
$$m = d^2/D^2$$
$\rho_1 =$ density at the upstream station
$\Delta p =$ differential pressure between stations 1 and 2

Equation (1.16) is used for differential pressure flowmeters with a coefficient of discharge, C, and, for compressible fluids, with an expansibility factor, ε, both of which we have omitted here. The full equation is given by equation (4.2).

Other equations relating to compressible flow will be introduced when we consider the critical flow venturi nozzle (cf. Baker 1996 for a fuller derivation).

1.7   FLUID PARAMETERS

We need to review briefly the main parameters used in flow measurement and list their units and typical values. Since the main concern of this book is with flow measurement, the reader is referred to other texts such as Noltingk (1988) for information beyond the following brief notes.

Temperature
The SI unit of temperature, T, is the degree Celsius (°C) or on an absolute scale the Kelvin (K). Table 1.3 compares these with the Fahrenheit scale (°F).

Types of thermometer in common use outside the laboratory are listed in Table 1.4. The choice of a thermometer will be influenced by whether or not an electrical output signal is required, the range, the accuracy required, convenience, and price. The platinum resistance thermometer (PRT), thermistor, and thermocouple provide a direct electrical output. It should be remembered that the thermocouple is essentially a temperature difference device and it may not always be convenient to retain one of the thermocouple junctions at, say, melting ice temperature. This is avoided in some commercial instruments by measuring the cold junction with a different sensor and correcting the reading. Ranges for various types are given in Table 1.4.

Table 1.3   Temperature scales

	°C	K	°F
Absolute zero	− 273.15	0	− 459.67
Ice point for water	0	273.15	32
Ambient range	10	283.15	50
	20	293.15	68
	30	303.15	86
	40	313.15	104
Boiling point for water at atmospheric pressure	100	373.15	212

Table 1.4   Thermometer ranges

Types of thermometer	Approximate range ($^\circ$C)
Expansion of solid	to +300
Mercury in glass	−40 to +650
Other liquids	−80 to +300
Gas	−130 to +540
Vapour	−50 to +320
Platinum resistance	−250 to +700
Thermistor	−100 to +300
Thermocouples	−200 to +1700
Optical pyrometer	+600 to +3000

Pressure

The SI unit of pressure is Newtons/metre^2 (N/m^2) or Pascals (Pa), but bar is commonly used as it is a more convenient size (cf. pounds per square inch, psi).

1 bar = 10^5 N/m^2 (approximately one atmosphere)

Subscript a is used for absolute pressure and g is used for gauge pressure

One atmosphere = 101 325 N/m^2
= 101 325 Pa
= 101.325 kPa
= 1.01325 bar
= 760 mm mercury column (Hg)
= 10.3 m water column (H$_2$O)
= 820 mm of water-over-mercury in a manometer
= 14.71 psi (lb/in^2)

Pressure is force per unit area. Pressure acts equally in all directions. It is useful on occasion to refer to pressure as a 'head'. The relationship between head and pressure is easily derived for a certain depth of liquid:

$$\text{Head} = \frac{p}{\rho g} \tag{1.17}$$

where ρ is the fluid density and g is the acceleration due to gravity.
Bernoulli's equation can be expressed in heads as

$$\frac{V_1^2}{2g} + \frac{p_1}{\rho g} + z_1 = \frac{V_2^2}{2g} + \frac{p_2}{\rho g} + z_2 \tag{1.18}$$

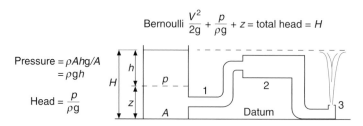

$$\text{Bernoulli} \quad \frac{V^2}{2g} + \frac{p}{\rho g} + z = \text{total head} = H$$

Pressure $= \rho Ahg/A$
$= \rho gh$

Head $= \dfrac{p}{\rho g}$

Fig. 1.8 Demonstration of variation in the terms of Bernoulli's equation (1.18). Note that this suggests an idealized flow without losses which could not be achieved in practice

Figure 1.8 shows how these three terms vary in an ideal loss-less flow system. At the tank surface, gauge pressure is zero and velocity is zero, so that $z = H$. At section 1, z is small, and so the terms in V and p will need to balance H. At section 2, the cross-section of the pipe is large, so that V is small, and the term in p will need to balance $H - z$. At section 3, gauge pressure is zero and z is small, so that the head H will virtually all be applied to $V^2/(2g)$. In practice losses would greatly reduce the height of the fountain.

If we apply equation (1.18) to the pitot and static tubes in Fig. 1.9, we find that at the mouth of the pitot, V is zero and the velocity can be obtained from the difference in the manometer column heights, h, as

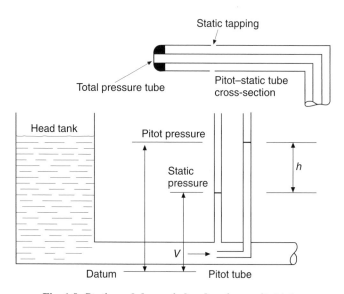

Fig. 1.9 Static and dynamic heads using a pitot tube

$$V = \sqrt{(2gh)} \qquad\qquad (1.19)$$

Pressure measurement is usually by one of three main methods (Baker 1996).

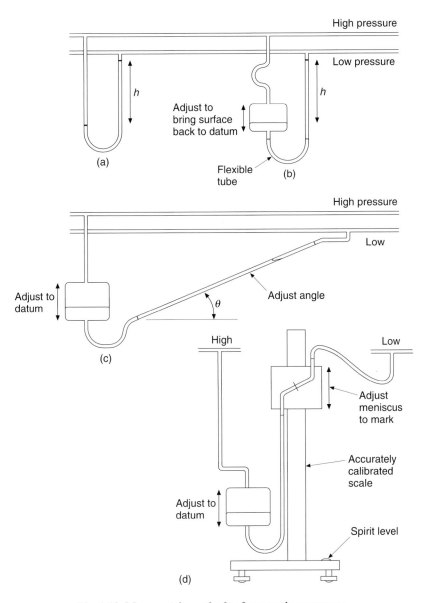

Fig. 1.10 Manometric methods of measuring pressure

1. A *manometer* which balances the pressure to be measured against a height of liquid (Fig. 1.10). This is the reason for giving the values of pressure in terms of mm mercury etc. Manometers are subject to errors, such as tube contamination, and depend on the precision of the tube bore and of the meniscus reading. A liquid giving an adequate displacement may not be compatible with a wide range of readings. Manometers with an electrical signal output are available, but expensive.

2. *Mechanical devices that deflect under pressure* such as the Bourdon tube (Figs 1.11a to c) which has a flattened tube wound into an arc. The tube unwinds under pressure since the flat cross-section becomes more oval and the increased distance between the arcs of inner and outer radii is

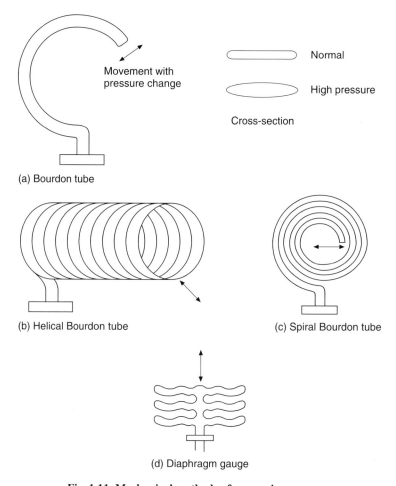

(a) Bourdon tube

(b) Helical Bourdon tube

(c) Spiral Bourdon tube

(d) Diaphragm gauge

Fig. 1.11 Mechanical methods of measuring pressure

accommodated by unwinding. A differential version requires the outside of the tube to be pressurized also. Another design uses diaphragms (Fig. 1.11d). The diaphragm elements are made up from pairs of corrugated discs with spacing rings welded at the central hole. Pressure will cause the assembly to elongate and this will in turn be used as a method of registering the pressure change.

3. *Electromechanical pressure transducers* (Fig. 1.12) are the most important and accurate devices for the future and are in a continual state of development. Some versions are:
   • capacitance transducers (Fig. 1.12a) in which deflection of one or two diaphragms causes movement of oil, which in turn deflects a third diaphragm which is a component in a capacitive element and the deflection is sensed as a change in capacitance;
   • piezo-resistive strain gauge (Fig. 1.12b) in which the deflection creates a change in resistance of an element;

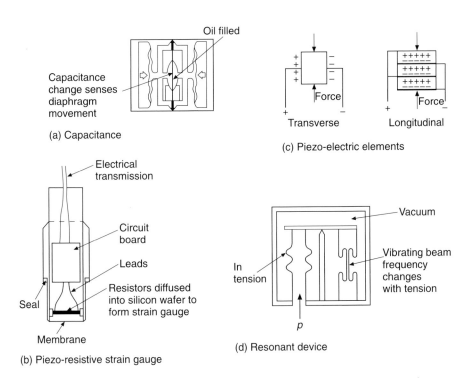

Fig. 1.12 Electromechanical methods of measuring pressure (after Higham in Noltingk 1988)

- piezo-electric (Fig. 1.12c) in which the pressure creates a voltage proportional to the pressure;
- resonant devices (Fig. 1.12d) in which the principle of a violin string is used, whereby the string resonates at a higher frequency for higher tension. Thus if the tension on the string can be changed by a deflection due to pressure change, then the frequency can be related to pressure.

Great care needs to be taken in designing the layout of pressure tappings and impulse tubes, and the arrangements of piezometer rings, which are manifolds allowing pressure tappings at several points around the circumference of the pipe.

Density

The SI unit of density, ρ, is kilograms/metre³ (kg/m³). The density of water at 0 °C is approximately

$$1000 \text{ kg/m}^3 = 1 \text{ g/cm}^3 \cong 62.4 \text{ lb/ft}^3$$

Specific volume is the reciprocal of density. Specific gravity is less used today, but it is defined as the ratio of the density of a material to the density of water at 60 °F. It is replaced by relative density:

$$= \frac{\text{liquid density at } T_1 \text{ °C(°F)}}{\text{density of water at } T_2 \text{ °C(°F)}}$$

Values of density, specific volume, and specific gravity are given in Table 1.5. The density of a liquid can be calculated from a knowledge of the mass of a known volume of the liquid. A gas density may be obtained in this way, but sensitivity of the mass balance needed to obtain a value of

Table 1.5   **Approximate values of density, specific volume, and specific gravity (values for gases at atmospheric pressure) (Kaye and Laby 1973, Hayward 1968)**

	Temperature (°C)	ρ (kg/m³)	v (m³/kg)	Specific gravity
Mercury	20	13 600	0.000 073 5	13.6
Water	20	1000	0.0010	1.0
Benzene	20	880	0.0011	0.88
Oxygen	20	1.31	0.76	0.0013
Nitrogen	20	1.15	0.87	0.0012
CO_2	20	1.80	0.56	0.0018
Air	20	1.18	0.85	0.0012

low uncertainty will be very high. Several commercial instruments are available which make use of a change in natural frequency of vibration, either of an element immersed in the liquid or gas, or of a pipe through which the liquid or gas flows.

Viscosity

Viscosity is quoted in two forms:

- *Dynamic (absolute) viscosity*, μ, is the ratio of shear stress to shear strain rate. The SI unit is the Pascal second (Pa s), but in common use is the centipoise (cP) where

    $1 \, cP = 10^{-3} \, Pa \, s$

- *Kinematic viscosity*, $v = \mu/\rho$, the ratio of dynamic viscosity to density. The SI unit is m²/s, but in common use is the centistoke (cSt) where

    $1 \, cSt = 10^{-6} \, m^2/s = 1 \, mm^2/s$

Some typical values of viscosity are given in Table 1.6.

1.8   MULTIPHASE FLOWS

The term multiphase flow is somewhat misleading as it covers both multi-component and multiphase. Thus dirty gas flows, air in water, cavitation, and wet steam may all, within this wide terminology, be termed two-phase flows. Three examples of such flows follow.

Table 1.6   Approximate values of density, and dynamic and kinematic viscosity (approximate values at 1 bar) (Kaye and Laby 1973, Hayward 1968)

			Viscosity	
	Temperature (°C)	*Density, (kg/m³)* ρ	*Dynamic, (cP)* μ	*Kinematic, (cSt)* v
Water	20	1000	1.00	1.00
Benzene	20	880	~ 0.65	~ 0.74
Oxygen	20	1.31	0.020	15.3
Nitrogen	20	1.15	0.018	15.6
CO_2	20	1.80	0.015	8.2
Air	20	1.18	0.018	15.4

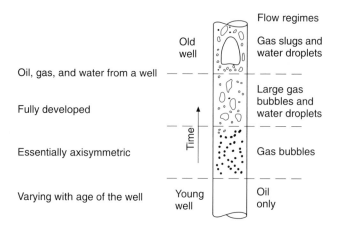

Fig. 1.13 An example of three-phase vertical flow: it should be noted that pressure change will alter the properties of components, and with time gas will come out of solution in the reservoir

Vertical multiphase flows

We consider first the flow in an oil well (Fig. 1.13). Although not strictly true, we assume that it is vertical, to indicate the general trends. When the flow reaches the well head it will have traversed a distance equivalent to as much as 30 000 pipe diameters. However, the flow may never reach a fully developed state, due to changing pressure and component ratios. Initially the flow will be single phase, essentially oil only. As the well pressure falls, gas bubbles will appear due to saturation of the oil; this is known as bubbly flow. With further ageing of the well, the gas bubbles may become larger. The size distribution will result from breakup due to turbulence, and coalescence due to the breakdown of the liquid film on close approach of two bubbles. However, water droplets are also now likely to be present, forming a third phase. Yet further ageing will result in the gas forming large slugs which travel up the centre of the pipe leaving a slower moving wall layer of liquid (which may even reverse direction during the passage of a bubble). These slugs may be up to 20 m long and might be much longer. They may form an equilibrium size distribution giving a balance between coalescence due to bubbles overtaking each other, and breakup due to instabilities when they become too large. Surrounding these bubbles will be smaller bubbles and droplets.

Horizontal two-phase flow

This is indicated in Fig. 1.14. The obvious difference is the loss of axisymmetry. Gravity now causes the less dense phase to migrate to the top of the

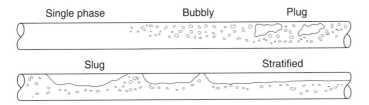

Fig. 1.14 Two-phase horizontal flow. Note that: fully developed flow may never be achieved; flow pattern is *not* axisymmetric; and to increase uniformity mixing may be used and a bubbly or droplet distribution will result temporarily but will quickly revert

pipe. Thus, in a gas/liquid flow the gas will move to the top of the pipe as bubbles. If these are allowed to become large, plugs of gas result, and as these coalesce slugs of gas take up regions against the top of the pipe. Eventually a sufficient number of these will lead to a stratified flow.

Alternatively the mixtures may be of two liquids such as water in oil. The droplets of water will sink towards the bottom of the pipe, mirroring the behaviour of air bubbles, and will eventually drop out on to the bottom of the pipe resulting in a continuous layer of water.

In most flows it will be difficult to predict the actual flow in such a situation and its effect on a flowmeter. It may be possible to mix the fluid to distribute the second phase, but this will cause severe turbulence and a changing profile, conditions generally considered unsuitable for flowmeter installation.

Steam

A truly two-phase fluid is wet steam. Superheated steam may be treated as a gas and its properties are well tabulated. However, it is of increasing importance to measure the flow of wet steam. This is steam made up of, say, about 95 per cent (by mass) vapour and about 5 per cent liquid. The droplets of liquid are carried by the vapour, but will not follow the vapour precisely. As with water droplets in oil, the liquid will drop through the vapour to impact on the pipe wall and we may obtain a liquid film until instability or turbulence causes a re-entrainment of this liquid film. The measurement of such flows poses major problems since the pressure and temperature remain constant while the dryness varies. We are, therefore, unable to deduce dryness or density from the pressure and temperature and as well as flowmeter errors caused by the wet steam, we shall not be able to obtain the mass flow from which the heat content is deduced. If the dryness is known a correction factor can be applied with caution (Miller 1996).

Other potential multiphase flows
Liquids can contain gases in solution. For water the maximum amount is about 2 per cent by volume. The gas in solution does not increase the volume of the liquid by an equal amount since the gas molecules 'fit' in the 'gaps' in the liquid molecular structures. For hydrocarbons the amount of gas which can be held in solution is very large and the gas–oil ratio (GOR) which is the volume of gas at standard conditions to the volume of liquid can range up to 100 or more. In either case, but particularly the latter, changes in flow conditions, for instance a pressure drop, can cause the gas to come out of solution resulting in a two-phase flow.

High humidity has problems of a not dissimilar type. If high humidity exists with a consequent large amount of water vapour, a change in conditions may result in the vapour changing to liquid droplets in the gas. This will also apply to other vapours.

Cavitation may occur in certain liquid flows at pressures around ambient. Cavitation is the creation of vapour cavities within the liquid due to localized 'boiling' at low pressure. It can cause damage since the cavities can collapse very quickly and erode a solid surface. It can also cause errors in readings, since it results in a larger volume than for the liquid alone.

Particulate matter can cause wear as well as errors, and may need to be removed with a fine filter.

CHAPTER 2

Calibration

2.1 DATUM CONDITIONS

Conditions for calibration:

- Known profile – upstream length of straight circular pipe sufficient to obtain fully developed flow (e.g. >60D).
- Known profile – appropriate Reynolds number (e.g. 10^6).
- Fluid as near to service fluid as possible.
- Smooth pipe upstream of flowmeter.
- Steady flow.
- Flowmeter mounted in a repeatable way.

The flow profile will affect the performance of most flowmeters. It is therefore essential that, for calibration, this is closely controlled. The above list contains those features which will cause the profile to be changed:

- in shape;
- in turbulence;
- in time.

We must, therefore, ensure that we calibrate a flowmeter with this datum. This will require that we set the flowmeter in a straight length of smooth pipe (the maximum permitted roughness will probably be in the range 10^{-3} to $10^{-4} D$) sufficient to generate a fully developed profile without swirl.

The fluid will need to be of a suitable type. This is particularly important with hydrocarbon flows where viscosity changes may introduce calibration changes. Unintentional second phases, such as air in water, can cause errors and should, therefore, be avoided.

Care should be taken over the concentricity of the flowmeters compared with the neighbouring pipework. Where high accuracy is sought it may be appropriate to position the flowmeter with spigots.

Flow steadiness is important since some meters are affected by unsteady flow and the unsteadiness may disturb the profile. Temperature and pressure variations can affect the viscosity of the fluid, as well as moving the calibration away from the datum. Conditions should, in general, be stable. Vibration is known to affect certain meters more than others and should be avoided.

2.2   STEADY FLOW

Errors due to unsteady flow result from:

1. Non-linear flowmeter characteristics.
2. Inertia of moving parts and of fluid.
3. Pulsation frequency close to operating frequency.
4. Secondary equipment unsuitable for unsteady signals.

The need for a steady flow, while expected, is particularly important for non-linear flowmeters. In practice, almost all flowmeters are affected by unsteady flows. This is due to four main reasons.

1. *Non-linear characteristic.* The differential flowmeter, which essentially uses equation (1.16) (with additional coefficients), has a quadratic characteristic, so that the mean pressure does not give the correct value of velocity. The same effect, but with more complex characteristics, may occur in other flowmeters.
2. *Inertia.* The inertia of a turbine wheel spinning in a gas flow will make it very difficult for such a flowmeter to follow the flow. It will tend to over-read. Again, similar effects occur in other flowmeters.
3. *Frequency of operation.* If the operating frequency of the flowmeter is close to that of the pulsations in the flow, then the flowmeter reading will be affected. An example of this is the vortex flowmeter, which has a tendency to 'lock on' to the pulsation frequency.
4. *Secondary equipment.* Most flowmeters have a secondary system which converts the flowmeter signal to a readable quantity. The manometer is one example of this which can be seriously affected by unsteady flows.

2.3   CALIBRATION RIGS FOR LIQUIDS

Flow calibration rig
*   Standing start and stop.
*   Flying start and stop.

For the linear flowmeter standing start and stop may be satisfactory, if integration of the flow is available.

Types of rig
Gravimetric –
*   mass of fluid in a measured time interval or total mass flowing through.

Volumetric –
*   volume of fluid in a measured time interval or total volume flowing through.

Having obtained a well-conditioned flow we need to measure it accurately. This requires either the measure of a certain volume passed in a measured time, or the mass passed in a measured time, or, for meters with sufficiently linear characteristics to give a total flow passed, the total volume or total mass may be sufficient, with flow stationary before and after.

The discussion which follows gives only a brief review of some of the main calibration methods. In particular, it does not attempt to review the methods in detail, as the reader who wishes to have a flowmeter calibrated will either send it to a recognized centre, or will seek advice on the construction of a suitable rig.

Figure 2.1 shows one example of a flying start and stop rig. Such rigs usually use water as the fluid. A diverter is used to switch the flow from the sump tank into a weighing tank for a timed period. Care is required in the design and adjustment of the diverter and timing switch, since it is virtually impossible to switch the flow from one tank to another instantaneously, and the timing switch must be adjusted to allow for the finite changeover time. The flow must be kept steady. This is traditionally achieved by a constant head tank, but is also commonly done with a closed loop with pump and sump tank. An interesting recent development is the use of a gyroscopic device to obtain the weight of the water. The

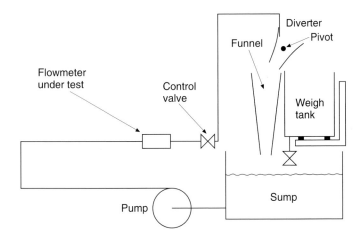

Fig. 2.1 Diagram of gravimetric flow circuit

precession of the gyroscope is directly related to the torque created by the weight of the tank. More common is the use of load cells, and in some cases a weighbridge may still be the most convenient method of weighing the water.

The uncertainty achieved in the measurement of water flowrate by national facilities, such as the UK's National Engineering Laboratory (NEL) and the US's National Institute of Standards and Technology (NIST), is in the range ±0.1 to 0.15 per cent.

The volumetric method is widely used and is most convenient in standing start and stop mode. A large vessel (Fig. 2.2) is calibrated, for example using weighed quantities of distilled water at a known temperature. The top and bottom of the vessel are of small diameter so that small volume changes cause measurable level changes compared with the large volume contained in the centre portion of the vessel. The initial fluid level is measured, often by means of a sight glass and weir, and the vessel is filled with the liquid being metered. The final level is arranged to be within the slender top section where it can be measured precisely with a surface point gauge or with a rule attached to the outside of the vessel. Hence, the volume which has flowed is known and the reading of the meter being calibrated can be compared with it.

Proving vessels of this type are used to check meter calibrations to uncertainties within ±0.1 per cent for custody transfer of hydrocarbon liquids. A typical vessel could be 500 litres capacity. Such vessels, of various sizes, have been used in calibration of meters for various liquids including liquid foods such as milk.

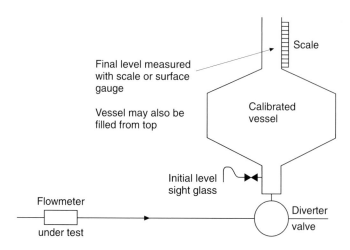

Fig. 2.2 Volumetric tank for liquids

Another type of volumetric rig makes use of the falling head method. A tall tank is filled with liquid. The liquid is allowed to flow through an outlet pipe to a flowmeter. The volume between two level switches is known and the meter is calibrated against this volume. The disadvantage of this method is that the flowrate will change slightly as the head falls.

The meter prover offers an alternative method of achieving a volumetric calibration by using the swept volume as a close-fitting sphere moves through a carefully constructed length of pipe between sensing switches. Figure 2.3 shows a unidirectional example. The sphere is projected into this pipe and its passage is recorded past two set points. The volume swept is known and the time taken is recorded. Other designs allow the sphere to pass both ways (bidirectional). Instruments of this type are commonly included in hydrocarbon metering stations.

This method has been widely used in the oil industry to provide on-site calibration, for instance, of banks of turbine flowmeters, the flow through which can be routed through the prover. The disadvantage of conventional provers is that they are very large and there are some applications where the space they take up may not be acceptable.

For reasons such as these, a new generation of compact provers has recently appeared. These devices still depend on the swept volume of a tube, but use a piston in a cylinder – a very much reduced 'pipe' length. With great manufacturing precision it appears that the performance of these compact provers can rival that of the large pipe prover. One problem that arises in the use of these compact provers is the small amount of fluid which passes. Thus, if a flowmeter produces a pulse for a certain volume of fluid passed, the number of pulses resulting from the fluid passed by a compact prover may be very low, causing a discrimination error. This has led to special techniques to overcome the problem, such as pulse interpolation.

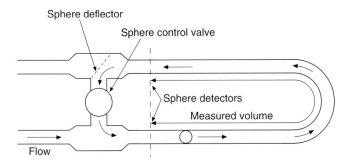

Fig. 2.3 Sphere prover

The conventional pipe prover may achieve uncertainties of the order of 0.1 per cent or less with calibration, and compact provers will seek to achieve similar performances if they are to compete.

2.4   CALIBRATION RIGS FOR GASES

- Bell prover – useful for manufacturers of gas meters such as those for domestic flows.
- PVT – assumes a known relationship for the gas.
- Gravimetric – requires great care in design and use.
- Soap film – for very low flows.
- Pipe prover – likely to become more widely used.

The bell prover, Fig. 2.4, is a volumetric calibration device for gases, which also uses standing start and stop. It is particularly suitable for small domestic gas meters and is capable of achieving an uncertainty of as low as ± 0.2 per cent if particular care is taken.

The PVT method assumes a known relationship between pressure, volume, and temperature for the gas. The gas is contained in a large calibrated vessel for which the pressure and the temperature are precisely known. The gas is allowed to flow through a heat exchanger, pressure control valve, and possibly a further regulator, before entering the meter under calibration. At the end of the calibration run, the volume of gas passed through the meter can be calculated from the change in the conditions in the calibrated vessel.

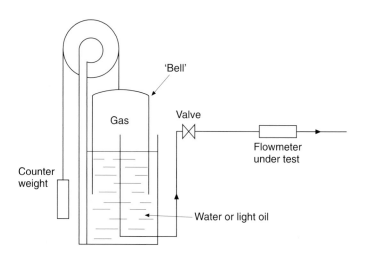

Fig. 2.4 Volumetric method for gases – bell prover

The gravimetric method, although most commonly used for liquids, is usable for gases if carefully designed and operated. However, it is a much more complex system than for liquids. A gas calibration rig using a gravimetric method is available at the National Engineering Laboratory (NEL), East Kilbride, Scotland. The uncertainty achieved for the UK's primary gas standard is in the range ±0.15 to 0.2 per cent.

For very low flowrates of gas a method making use of the movement of a soap film through a vertically mounted burette is sometimes used.

Although the pipe prover does not appear to have been applied to the calibration of gas meters, the compact prover offers an option for gases and may come into wider use in the future.

2.5 MASTER METERS

Master meters

- Very high accuracy.
- Preferably operating in pairs to cross-check.
- Preferably calibrated with upstream pipework and flow straightener.
- Providing a transfer standard from a nationally certified rig.

A further option, of particular use for an organization needing traceable calibrations, but unable to justify a full calibration rig, is to make use of a transfer standard flowmeter against which to calibrate their flowmeters. This requires a master meter of high accuracy and stable operation. Calibrated master meters may be used to measure the flow in a pipe and, hence, to calibrate other meters. This may be the most economical option for some organizations. Positive displacement meters are common, but turbine and other meters have been used. For gases, sonic nozzles provide a satisfactory method, and wet gas meters are suitable due to their high performance. These instruments will be described later.

To achieve a check on the performance of a master meter they are often used in pairs, either in series, so that the consistency of their readings is continually checked, or in parallel when one is used most of the time and the second is kept as a particularly high accuracy meter for occasional checks.

To achieve very high performance the meter is calibrated with the upstream pipework and a flow straightener permanently in position, and the use of two of these packages in series. In this way any deviation caused by installation or by drift of one meter will appear as a relative shift in calibration between the two meters. Such meters or meter pairs are referred to as transfer standards, in that they allow a calibration standard to be transferred (with only a small increase in uncertainty) from a nationally certified rig to a manufacturer's rig or a research laboratory rig.

2.6   SITE CALIBRATIONS

Proving vessels
Pipe provers
Compact provers
Reference meters
Clamp-on meters
Probes
Tracers
Inspection and measurement
Self-checking systems

Site (*in situ*) calibration is valuable, in that it takes account of installation effects, but in some cases it is very difficult to achieve with a small value of uncertainty. In such cases, therefore, it may be preferable to refer to it as *site verification*.

Proving vessels, pipe provers, compact provers, and reference meters may be suitable for use in many petroleum product installations, where suitable connections are available. Uncertainties in the order of 0.2 per cent and better are achievable.

Clamp-on meters (ultrasonic), probes, and tracers may be used. However, they require considerable expertise and care to achieve a calibration which is unlikely to be better than 1 per cent and, in many cases, it may be difficult to achieve better than 5 per cent.

Inspection and dimensional measurement of installed flowmeters depend either on standard instruments (differential pressure) or good historical documentation from original calibration and installation.

Some systems may be designed as self-checking either by redirecting the flow to a second (master) meter or by using flowmeters with a self-checking function.

2.7   GENERAL COMMENTS

For calibration to be acceptable, the ultimate source of the measurement must be known and the calibration must be traceable to that standard. Thus, in the UK, the national standards for flow measurement are held by the NEL. These standards are, themselves, traceable back to more fundamental measures of mass, time, and volume. As a result of this, a traceability chain is formed. Each link is formed from a rig or flowmeter calibrated against a rig or flowmeter of greater accuracy, with sufficient frequency to ensure continuing confidence. It is clear, therefore, that if a transfer standard with an uncertainty of, say, ±0.25 per cent is used to calibrate a flowmeter, the uncertainty in the calibration of that flowmeter will be greater than ±0.25 per cent.

For a standing start and stop gravimetric rig, only one measurement, mass, needs to be traceable, but for a flying start and stop, time must also be traceable. However, for standing start and stop, the behaviour of the meters at low flowrates near start and stop may prejudice the calibration.

It can be seen from the discussion above that the calibration uncertainties achievable at present are of the order of:

For liquids: ±0.1 to 0.15 per cent
For gases: ±0.15 to 0.25 per cent

Where total mass or volume, rather than flowrate, is required, these values may be marginally improved.

CHAPTER 3

Selection

3.1 CONSIDERATIONS IN SELECTING A FLOWMETER

We shall consider:

- type of fluid;
- special fluid constraints;
- flowmeter constraints;
- environmental considerations;
- total cost.

For the next few chapters we shall be discussing the operation, advantages, and disadvantages of the most common meters available commercially or (in the case of the differential pressure meters) able to be manufactured by a prospective user. Some readers may read later chapters for general interest. Most will probably have a specific application in mind and will want to be guided to the instruments most appropriate to it. This chapter attempts to guide, recognizing that the uninitiated will be looking for sufficient knowledge to deal intelligently with manufacturers and their brochures, and that the initiated will probably have bypassed this introductory book.

In the limited scope of this text it is not possible to be exhaustive about the many facets of flow measurement and flowmeters which one needs to take into account in making a selection. An attempt has been made here to focus on the more general considerations leaving the particular to the manufacturer to supply relative to his design.

The selection table (Table 3.1) concentrates on the main types of meter. For convenience they are subdivided into momentum, volume, and mass sensing meters. This subdivision also emphasizes that, for instance, a meter which responds to momentum will be sensitive to density change. Where a less common meter of the same family has particular features which make it unique or particularly useful in certain applications, these may be indicated in notes in the table and in the discussion in subsequent chapters, where other points arising from this chapter will be elaborated.

Table 3.1 Flowmeter selection summary table (from Baker 2000 reproduced by kind permission of Cambridge University Press)

1 Flowmeter type[a]	2 Liquid or gas	3 Slurry	4 Other two-phase	5 Accuracy	6 Typical turndown[d]	7 Diameter range (mm)[e]	8 Temperature range (°C)[f]	9 Flow range $[m^3/h(kg/h)]$[f]	10 Pressure loss	11 Sensitivity to installation[g]	12 Initial cost	Notes
Momentum[a]												
Orifice	L	X	?	**	5:1	50–1000		$1–3 \times 10^5$	H	H–L	L/M	Concentric ISO orifice with differential pressure cell assumed
	G	X	?	**	5:1	50–1000		$10–4 \times 10^6$				
Venturi	L	*	*	**	5:1	50–1200		30–7000	M	L	M/H	
	G			**	5:1	50–1200		$400–10^5$				
Nozzles	L	?	*	**	5:1	50–630		$2–1.7 \times 10^4$	M	H–L	M	Does not include critical nozzle
	G			**	5:1	50–630		$20–2.5 \times 10^5$				
Variable-area	L	X	X		10:1	15–150	−200 to 350	$10^{-1}–100$	M	L/M	L	Glass/plastic assumed: higher ratings for steel
	G				10:1	15–150		$10^{-2}–2000$				
Other venturi-like	L	?	?	?	3:1	13–1200			M	H/M	M/H	Consult manufacturers
	G				3:1	13–1200						
Averaging pitots	L	X	X	*	10:1	25–12 000	up to 450	$10–3 \times 10^4$	L	H	M/L	Installed accuracy possibly lower for insertion designs
	G			*	10:1	25–12 000	−100 to 450	$200–6 \times 10^5$				
Laminar	G	X	X	?	20:1	?		up to 120	H	N	M	Particularly appropriate for pulsating flows
Volume												
Positive displacement	L	X[b]	X[b]	***	10:1	4–200	−50 to 290	0.01–2000	H/M	N	H/M	Turndown given may be achieved for rotary positive displacement gas meter
	G			***	80:1		−40 to 65	0.01–1200				
Turbine	L	X	X	***	10:1	5–600	−265 to 310	0.03–7000	M	M/H	L/M	High precision instruments assumed rather than water meters etc.
	G			**	30:1	25–600	−10 to 50	0.01–25 000				

Type	Fluid			Ratio	Size (mm)	Temp (°C)	Flow range				Comments
Vortex	L	X	**	10:1	12–200	−200 to 400	3–2000	H/M	M	L/M	Fluidic flowmeter suitable for wide range of utility flows
	G	X	**	10:1	12–200		50–10⁴			L/M	
Electro-magnetic	L	**	**	10:1 (or greater)	2–3000	−50 to 190	10^{-2}–10^5	L	M/L	L/M	Only available commercially for conducting liquids
Time-of-flight ultrasonic	L	?^b	?^b	20:1	10–2000	−200 to 260	3–10^5	L	M/L	M/H	Single beam are more sensitive to installation
	G		**	30:1	20–1000	−50 to 260	0.04–10^5		M/L	M/H	Correlation and Doppler for two-phase flows
Mass											
Multisensor	L	*^c	*^c	?	?	?		H		H	
	L	**^c	**^c	?	?	?		?		H	
Wheatstone bridge	L	X	**	50:1	6–60	−18 to 150	$(0.05$–$2.3 \times 10^4)$	?	L	H	Specialist instrument especially for engine testing
Thermal	L	X	*	15:1	2–6	0 to 65	$(0.002$–$100)$	M	L (CTMF) M/H (ITMF)	M	Higher flows require CTMF with bypass or ITMF
	G		*	50:1	6–200	−50 to 300	$(2 \times 10^{-4}$–$8000)$	L/M		M	
Angular momentum	L	X	*	7:1	20–50	−40 to 150	$(100$–$4.5 \times 10^3)$	M	M	M	Particularly suitable for aircraft fuel flow
Coriolis	L	***	X	100:1	6–200	−240 to 200	$(1$–$7 \times 10^5)$	M/L	N	M/H	Straight single tube has essentially zero pressure drop; gas ranges are probably more limited
	G	**	X								

Key: ***, very high; **, more suitable/high; *, suitable/medium; ?/blank, uncertain/lower; X, unsuitable; N, negligible; L, low; M, medium; H, high.

a Some proprietary devices offer special features: higher differential pressure, linear characteristics.

b Some designs have been produced for slurries.

c Multisensor systems have been developed specially for multiphase.

d Typical estimates that may be exceeded by using smart transducers and may be greater or less than the value attained by some designs.

e Larger or smaller sizes may be available.

f Other ranges may be available.

g Flow conditioning may be used in some applications to reduce this effect.

CTMF, capillary thermal mass flowmeter; ITMF, insertion and in-line thermal mass flowmeters.

3.2   NATURE OF THE FLUID TO BE METERED

Type of fluid:

- liquid or gas;
- slurry;
- multiphase.

The first choice, liquid or gas, is the most obvious, and will cause a few types of meter to be eliminated, although most of the main types have designs suitable for one or the other. Increasingly, there is interest in the measurement of slurries and two- (or more) phase flows. The selection of suitable meters for these is restricted at present. In the table 'X' indicates unsuitable and '*' indicates suitable except where some are more suitable '**' than others. '?' indicates that it is not common, but might be appropriate, in this application.

The result is rather subjective, as will become apparent as each meter is considered. For instance, the electromagnetic flowmeter has found a valuable role in slurry flow measurement and is almost unique in this. However, the Coriolis mass flowmeter is claimed to be suitable for slurry, and an ultrasonic flowmeter has been specially designed for this purpose. The electromagnetic flowmeter is also useful in two-phase flows when the continuous phase is conducting. However, venturis are usable, but with caution.

A note in the table indicates that the ultrasonic cross-correlation flowmeter and the Doppler flow monitor may also be worth considering in two-phase flows. However, they both suffer from problems relating to what is being sensed, and this may limit uncertainty to, at best, ±2 to 3 per cent.

Special fluid constraints

- Clean or dirty.
- Hygienic.
- Corrosive.
- Abrasive.
- High flammability.
- Low lubricity.
- Fluids causing scaling.

In addition, and after initial selection of the most likely types, it will be necessary to discuss special fluids with manufacturers.

Resulting constraints will vary according to materials and design details and will not necessarily be related to a particular flowmeter design. However, flowmeters with rotating parts will be less suitable for dirty, abrasive, or low lubricity fluids. It is worth noting that the term *low lubricity* appears to have been adopted to indicate a situation where a high friction coefficient occurs in a bearing with low viscosity fluid acting as lubricant (Baker 2000). Abrasion may also detract from the performance of other meters, such as the vortex meter.

Another example of a fluid constraint is that the liquid must be conducting for the application of current commercial electromagnetic flowmeters.

3.3 FLOWMETER CONSTRAINTS

Flowmeter constraints

- Accuracy (or measurement uncertainty after calibration).
- Diameter range.
- Temperature range.
- Pressure maximum.
- Viscosity range.
- Flow range.
- Pressure loss created by the flowmeter.
- Sensitivity to installation.
- Sensitivity to pipework supports.
- Sensitivity to pulsation.
- Whether the flowmeter has a clear bore.
- Whether a clamp-on version is available.
- Response time.
- Ambient conditions.

Apart from the initial cost of the flowmeter, which should be considered last, the main considerations will be as shown in the box above.

Accuracy is perhaps the most difficult to determine, since both user and manufacturer are prone to overstate their requirements and capabilities. In the table:

*** is for measurement uncertainty after calibration better than 0.5 per cent of rate.

** is for measurement uncertainty after calibration better than 1 per cent of rate.

* is for measurement uncertainty after calibration better than 2 per cent of rate.

? is for measurement uncertainty unknown.

The estimates are the author's view of the accuracy of flowmeters indicated by their measurement uncertainty after calibration. These estimates are based on the likely random error inherent in the operating principle of the flowmeter. They should be achievable with a high quality example of the flowmeter. For a standard orifice the uncertainty before calibration due to the discharge coefficient and other parameters is likely to be 1.0 to 1.5 per cent. After calibration the total uncertainty will be less than this, and it is reasonable to assume that 0.5 per cent could be achieved. This appears to be confirmed by data. Some manufacturers may feel able to claim higher values, but the reader should ask for justification for such claims. For instance, some manufacturers claim that a single beam ultrasonic flowmeter is capable of a measurement uncertainty of less than 0.5 per cent. However, this performance needs adequate substantiation before being accepted. Several other commercial flowmeter designs have claims of better than 0.5 per cent, but again this should be viewed with scepticism, since it indicates a level of accuracy which is very hard to achieve. Some manufacturers may not always achieve the values indicated. The measurement uncertainty of a variable area meter is often quoted as 1–2 per cent of full scale. On balance it is taken here as just outside the 2 per cent of rate category.

Measurement uncertainty should be checked regularly by recalibration since changes in bias, or systematic, error may occur due to changes in method of measurement, location, or long elapsed times. It may be wise to recalibrate after a 6-month period and, on the basis of the measured shift, to identify the period for succeeding calibrations.

Diameter, temperature, and flow are the author's best estimates based on manufacturers' literature, and the general references at the end of the book. In all cases the diligent reader can probably find examples of greater ranges than those given, and manufacturers are always seeking to extend the capability of their instruments. Note that in the column for flow range, the values in parentheses indicate kg/h rather than m³/h. Table 1.1 has a conversion in terms of water flow.

The maximum pressure has not been included since, apart from exceptionally high pressure applications, when one might turn to an orifice plate or a turbine meter, there appear to be versions of most designs which are offered for normal industrial ranges. In some cases minimum pressures may be important, as cavitation may be a danger, and the manufacturer should be consulted.

The pressure loss is given as 'H' (high), 'M' (medium), or 'L' (low). Only the electromagnetic and ultrasonic meters have been categorized as L, and for each there is virtually no more pressure loss than in a similar length of pipe. The orifice plate and target are H, and positive displacement (PD) and vortex are H/M, although their relative losses will depend on such things as the instrumentation being driven by the PD meter. Earlier twin-tube Coriolis meters had a greater pressure loss than the new generation of single straight-tube Coriolis meters which should have no more loss than the straight tube they replace. The Coriolis meter has, therefore, been rated M/L.

The positive displacement meter may be assumed to be insensitive to upstream installation effects and is, therefore, rated N. Most other flowmeters are affected and the values in Table 3.1, which should be read in conjunction with Table 3.2, are provided as an approximate guide to the required upstream and downstream spacing for various fittings.

All of the categorizations are best estimates by the author, based on a cautious interpretation of published data and manufacturers' information, although some manufacturers may be prepared to uprate them. The values for valves have been omitted from Table 3.2, since there are too many uncertainties of valve type and opening and, in any case, it is not recommended to place a meter downstream (say, less than 60D) of a valve.

The behaviour of variable area and Coriolis meters is based on the assumption that the flow direction change normally associated with their inlet piping will reduce sensitivity to other upstream fittings, but there is some suggestion that the Coriolis meters are not entirely unaffected by inlet flow profile, and 5D has been recommended for the variable area meter.

Data on pulsation effects are too inadequate to make any useful categorization. Most flowmeters are sensitive to pulsation over part of their range, with the possible exception of the ultrasonic flowmeter and also the viscous flowmeter, which was specially designed to cope with pulsating gas flows.

The flowmeters with essentially a clear bore (non-intrusive) are the electromagnetic, ultrasonic, and single straight-tube Coriolis types. The only flowmeter which may fail and block the line is the positive displacement flowmeter. All the others have a partial line blockage, although the venturi has such a smooth change in section that its effect is probably of small concern.

Only the ultrasonic flowmeter family offers clamp-on options (non-invasive).

If a high speed of response is required this may eliminate some designs (for instance, for control installations where a response in less than a second may be required).

Table 3.2 An approximate sensitivity rating for upstream and downstream installation spacing in diameters (D) that may be required to retain claimed accuracy for flowmeters of various types. Levels of sensitivity: H, M, L and N are suggested in Table 3.1 on the basis of manufacturers' information and published data (this table is adapted from ISO 5167)

	Upstream spacing between specified pipe fitting and flowmeter					*Down-stream spacing after meter*
Sensitivity of flowmeter to installation	*90° bend or tee*	*Two bends same plane*	*Two bends perpendicular planes*	*Expander 0.5D to D over D to 2D*	*Reducer 2D to D over 1.5 to 3D*	
Negligible (N)	No additional upstream or downstream pipework needed before meter					
Low (L)	10	16	34	16	5	4
Medium (M)	14–18	18–26	36–48	16–22	5–9	6–7
High (H)	28–36	36–42	62–70	30–54	14–22	7–8

Finally the compatibility of the flowmeter and its related instrumentation to the ambient conditions should be checked with the manufacturer.

3.4   ENVIRONMENT

Environmental considerations

- Ambient temperature.
- Humidity.
- Exposure to weather etc.
- Level of electromagnetic radiation.
- Vibration.
- Tamper proof for domestic use.
- Classification of the area requiring explosion proof, intrinsic safety, etc.

The installation of the flowmeter will also be affected by environmental considerations. These will need to be assessed in consultation with particular manufacturers. Virtually all commercial designs are capable of withstanding or meeting all these environmental considerations and most manufacturers will offer particular versions to cope with particular constraints. One apparent exception to this (at present) is vibration which may affect, for example, some types of Coriolis meters.

In this book, I have not attempted to provide guidance on special application design such as that for domestic use or on safety and other requirements. Other books such as that produced by Endress+Hauser (1989) give information on such matters.

3.5   SPECIAL EFFECTS

Under certain circumstances cavitation can occur in a liquid, gas or air may be entrained in the liquid, and droplets can form in a saturated vapour. In addition, some flows may be particularly abrasive. These needs should again be assessed in conjunction with the manufacturer, or with experts from whom advice may be sought.

3.6   PRICE

This is possibly the most important consideration and is difficult to tabulate since the range of price for any design is wide and the datum is always changing! The author has, therefore, attempted to rate the initial cost of the flowmeters as 'H' (high), 'M' (medium), or 'L' (low). However, the medium bracket tends to become a 'catch-all' and covers, for instance, electromagnetic, turbine, and vortex flowmeters, which can actually have an initial cost range of 4:1 for a 100 mm diameter design. For this reason L/M or M/H have been used where low or high cost versions are available.

3.7   CHOOSING

Although there may not be a true expert system for flowmeter selection, there are now software selection tools, in many cases produced by manufacturers, that allow a user to select a specific meter type against a set of conditions. These tools do not replace the metering specialist, but they do allow suitable meters for the more straightforward applications to be specified.

The object of a book such as this, or the other books listed in the bibliography, some of which are very detailed, is to aid the user in making the initial selection, and to educate him or her sufficiently to be able to ask the manufacturer the right questions. If, say, a mass flowmeter is definitely required, then the selection is reduced. Even so, the option of combining a momentum-, or a volume-sensitive flowmeter with a density cell should not be overlooked, and may be an option which offers greater confidence.

The first column of Table 3.1 is a partial list of the types of flowmeter available, and the next three columns can be used to eliminate those which are clearly inapplicable.

Column 5 will, in most cases, only allow one or two instruments to be eliminated unless very high accuracy is required.

Column 6 gives likely turndown for those applications where there is a requirement for a range of flows.

Columns 7 to 9 can be used to eliminate further by excluding those flowmeters which do not meet the requirements of the particular application. Again for most applications this will not eliminate many flowmeters.

Column 10 may be a critical consideration in a few applications, but since orifice plates with high pressure loss dominate industrial flow measurement, it is unlikely that pressure loss will be an overriding consideration. However, with greater energy conservation this may influence the final choice.

Column 11 may be of greater importance, as in many applications the installation options are limited and the meter may have to be installed in a very short length of straight pipe. In this situation upstream pipe fittings may affect the meter reading. If a meter with high sensitivity to flow profile distortion has to be used, some improvement may be achieved by installing a flow conditioner, but this must be done with care, may add to the pressure drop in the line, and may cause an unacceptable obstruction.

Column 12 can be no more than an indication of broad ranges of initial cost. Thus, if this is a dominant factor, and budget is restricted, then a variable area device may be the only option. There are other aspects of cost which have not been mentioned. Initial cost will, of course, include both purchase price and costs of installation. There will also be ongoing costs associated with maintenance, energy loss due to the presence of the flowmeter, and savings due to the information which the flowmeter provides. If the flowmeter is clearly required, then the savings should outweigh the costs. This does not reduce the responsibility of the user to install the best and most economical instrument for the job!

CHAPTER 4

Momentum Flowmeters

Standard instruments
- Orifice plate.
- Venturi meter.
- Nozzles.
- Specific orifice plates.
- Sonic nozzle.

Other pressure-sensing devices
- Proprietary designs.

Other momentum-sensing devices
- Drag plate/target meter.
- Rotameter/variable area meter/float-in-tube.

In this chapter we consider flowmeters which sense the momentum of the flow, the most important of which are the differential pressure (DP) devices. It is appropriate to take the differential pressure flowmeters first, since they have a long and distinguished history and will continue to dominate the flow measurement scene despite many alternative methods. The reasons for this dominance can be attributed to the following factors.

(a) a traditional conservatism which trusts the known device, despite its shortcomings, rather than the unknown device;
(b) a wealth of experience which has been distilled into BS EN ISO 5167 allowing a flowmeter of known measurement uncertainty to be designed and constructed by a well-founded flow laboratory;
(c) a mistaken view that it is a simple and cheap matter to design and build a flowmeter of 'high accuracy'. It is not, but that is a later part of the story.

The flow through the venturi, see Fig. 4.1, illustrates the well-behaved nature of a flow when the area changes blend reasonably smoothly from one size to another. Converging flow is particularly well behaved, while diverging flow for diffusion, with small enough angle, will continue to remain generally in one direction. As the velocity increases due to the smaller area, so the pressure decreases, and vice versa. However, 'pressure recovery' is dependent on the smoothness of the expanding flow.

In contrast, the flow through the orifice plate is far from smooth, see Fig. 4.1. The abrupt area change caused by the orifice plate, while causing the

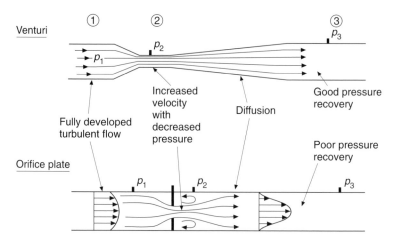

Fig. 4.1 Differential pressure flowmeters

flow to contract as it passes through the orifice, may also cause a small recirculation zone in the corner upstream of the plate and will cause a large recirculation zone downstream of the plate around the central jet area. In addition to this the flow downstream of the plate is highly disturbed and the diffusion will result in high pressure losses.

The venturi was invented by Clemens Herschel (1842–1930), a graduate of Harvard University, and he named it after Venturi in recognition of his important research on the device. Physically it illustrates some important ideas of the differential pressure device. Because it has a tapered inlet converging from station 1 to station 2 (Fig. 4.1) the losses in the flow are very small and equation (1.16) gives a good estimate of the velocity in the main pipe based on the pressure difference, Δp.

$$V_1 = \frac{m}{\sqrt{(1-m^2)}} \left(\frac{2\Delta p}{\rho} \right)^{1/2} \tag{4.1}$$

where $m = d^2/D^2$, D is the main pipe diameter, and d is the throat diameter. By using a tapered diffuser the flow is caused to diffuse with less loss than results from a sudden expansion, but there will still be some loss of total pressure between p_1 and p_3. We will return to the venturi meter after considering the orifice meter – the most common of the differential-pressure (DP) meters.

However, it is appropriate to note at this point that ρ, the density, appears in the equation, confirming that, since these meters sense momentum, it is necessary also to obtain a value of the density.

4.1   ORIFICE PLATE METER

The orifice plate flowmeter is the most common industrial instrument. It is apparently simple to construct and has a great weight of experience to confirm its operation. However, it causes a complex flow pattern compared with the smooth venturi flow, Fig. 4.2.

The inlet flow will almost always be fully developed turbulent if adequate upstream pipe length is allowed and the international and national standards are followed (BS EN ISO 5167). In the following description the design with pressure tappings at one diameter, D, upstream of the orifice plate, and half a diameter, $D/2$, downstream is considered. The other standard arrangements are flange and corner tappings. In the D–$D/2$ arrangement the upstream pressure tapping senses the pressure essentially before any redistribution of the flow occurs. This flow converges to the orifice and continues to converge due to the inward momentum created, to the *vena contracta*, the narrowest point of the submerged jet, downstream of the orifice. Outside this position is a recirculation region across which the pressure is sensed at the wall. The flow downstream of this point is disturbed and pressure recovery is much less

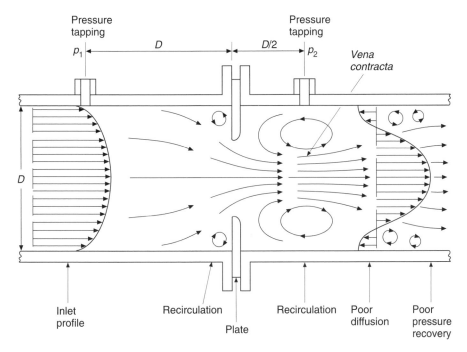

Fig. 4.2 Diagram to show geometry and flow patterns in the orifice plate flowmeter (from Baker 2000 reproduced by kind permission of Cambridge University Press)

efficient than in the venturi meter. The resulting pressure loss across the flowmeter is much greater than for the venturi meter.

Were we able to measure the diameter of this *vena contracta*, equation (1.16) could be used as a reasonable approximation to the relationship between flowrate and pressure drop. Recognizing that this is not easily obtainable, a discharge coefficient is used.

The nature of the flow suggests why the orifice plate performance is particularly sensitive to certain design details. Thus, the sharpness of the plate leading edge is likely to affect the development of the *vena contracta*. (Note that the standard orifice plate is bevelled, if necessary, on its downstream side.)

Because of the losses in the actual flow, equation (1.16) is modified by the addition of coefficients which allow for the divergence from ideal incompressible flow. For mass flowrate the differential pressure flowmeter equation becomes

$$q_{\mathrm{m}} = CE\varepsilon_1 (\pi/4)d^2 \sqrt{(2\rho_1 \Delta p)} \tag{4.2}$$

where

$E = 1 \big/ \sqrt{(1-\beta^4)} = 1 \big/ \sqrt{(1-m^2)}$ where $\beta = d/D$ and $m = d^2/D^2$

C = coefficient of discharge

ε_1 = expansibility factor based on the pressure at the upstream tapping point (= 1 for incompressible fluids and allows a correction where a flowmeter for gas is calibrated with water)

d = orifice diameter

D = upstream pipe diameter

Δp = differential pressure

ρ_1 = density at the upstream pressure tapping

The differential pressure flowmeter equation is given in BS EN ISO 5167, and an expression is given for the coefficient of discharge, C. The coefficient allows for the very different flow in the orifice plate as compared with an ideal contracting flow and, in particular, allows for the size of the *vena contracta* compared with the orifice.

The form of the coefficient of discharge is likely to be the provisional form of the equation as set out in ISO 5167-1: Amd 1: 1998, and by Reader-Harris and Sattary (1996).

The equation for C, which has come to be known as the Reader-Harris/Gallagher equation, is as follows.

For $Re \geq 4000$ and $D \geq 71.12$ mm (2.8 in):

$$C = 0.5961 + 0.0261\beta^2 - 0.216\beta^8 \qquad C_\infty \text{ term}$$

$$\left.\begin{array}{l} + 0.000521\left(\dfrac{10^6\beta}{Re}\right)^{0.7} \\[4mm] + (0.0188 + 0.0063A)\beta^{3.5}\left(\dfrac{10^6}{Re}\right)^{0.3} \end{array}\right\} \quad \text{Slope term}$$

$$\left.\begin{array}{l} + \left(0.043 + 0.080e^{-10L_1} - 0.123e^{-7L_1}\right) \\[2mm] \times (1 - 0.11A)\beta^4\left(1 - \beta^4\right)^{-1} \end{array}\right\} \quad \text{Upstream tap term}$$

$$- 0.031(M_2' - 0.8M_2'^{1.1})\beta^{1.3} \qquad \text{Downstream tap term}$$

Where $D < 71.12$ mm (2.8 in) the following term should be added:

$$+ 0.011\,(0.75 - \beta)(2.8 - D/25.4) \qquad (D \text{ in mm}) \qquad (4.3)$$

where Re is based on the pipe diameter D,

$$M_2' = 2L_2'/(1 - \beta) \qquad A = (19\,000\beta/Re)^{0.8}$$

where $L_1 = l_1/D$ and l_1 is the distance of the upstream tapping from the upstream face of the plate, and $L_2' = l_2'/D$ and l_2' is the distance of the downstream tapping from the downstream face of the plate. The prime signifies that the measurement is from the downstream and not the upstream face of the plate.

The terms in A are only significant for small-throat Reynolds number. M_2' is in fact the distance between the downstream tapping and the downstream face of the plate divided by the dam height.

For the purposes of equation (4.3) the values of the upstream and downstream lengths are given below:

	L_1	L_2'	
D and D/2	1	0.47	
Flange	25.4/D	25.4/D	(D in mm)
Corner	0	0	

The uncertainty associated with equation (4.3) is:

$(0.7 - \beta)\%$	for	$0.1 \leq \beta < 0.2$
0.5%	for	$0.2 \leq \beta \leq 0.6$
$(1.667\beta - 0.5)\%$	for	$0.6 < \beta \leq 0.75$

and for $D < 71.12$ mm (2.8 in) the following should be added arithmetically:

$$+ 0.9(0.75 - \beta)(2.8 - D/25.4) \qquad (D \text{ in mm})$$

The previous expression which appeared in earlier versions of ISO 5167 was known as the Stolz equation, and was rather simpler than this one. Whichever equation is used, it is important to ensure that the specified constraints are met on d, D, β, and Re. Outside these limits the standards give no authority for its use. Within these limits the value of C is said to have an uncertainty of at best 0.5 per cent.

The expansibility factor is briefly defined above and expressions for this are of the form

$$\varepsilon_1 = 1 - (a_\varepsilon + b_\varepsilon \beta^4) \frac{\Delta p}{\kappa p_1} \tag{4.4}$$

where $a_\varepsilon = 0.41$ and $b_\varepsilon = 0.35$ are constant coefficients used in ISO 5167-1: 1991, κ is the isentropic exponent, which for an ideal gas is equal to the ratio of the specific heats. p_1 is the pressure at the upstream tapping. In ISO 5167-2: 2002 the equation may be

$$\varepsilon_1 = 1 - (0.351 + 0.256\beta^4 + 0.93\beta^8) \left[1 - \left(\frac{p_2}{p_1} \right)^{1/\kappa} \right] \tag{4.5}$$

which is of the same form as equation (4.4) if β^4 and $\Delta p / \kappa p_1$ are both small.

The pressure loss across the meter is given by various expressions in ISO 5167, but a recent version of the standard allows

$$\text{Pressure loss} \approx (1 - \beta^{1.9}) \Delta p \tag{4.6}$$

This suggests that for $\beta = 0.5$, less than 30 per cent of the differential pressure, Δp, through the orifice is recovered in the downstream diffusion.

It is important to appreciate that the constraints set out in the standards must be respected if the uncertainty values are to be used with any authority. It is also important to realize that when combined with the other uncertainties in equation (4.2), the best overall uncertainty which can be achieved for the orifice plate is likely to be of the order of 1.5 per cent, and could be worse than this. Examples of manufacturing and installation constraints on the plate are given in Table 4.2. The orifice is not a simple device from which high accuracy is automatically obtained. The attainment of high accuracy is only as a result of careful observation of the correct design procedures and manufacturing requirements. The reader should check the latest version of ISO 5167 for changes in recommended installation lengths.

Table 4.1   Orifice plate manufacture, installation, and uncertainty

Examples of manufacturing constraints:
  Flat to within 1%
  Plate thickness to within $0.001D$
  Upstream edge radius $\leq 0.0004d$
  Downstream $(D/2)$ pressure tapping position for $\beta \leq 0.6$ to $\pm 0.02D$

Examples of installation constraints (add 0.5% uncertainty for parenthesized numbers):

		Upstream diameters		*Downstream diameters*
β	*90° bend*	*Two 90° bends in perpendicular planes*	*Gate valve fully open*	*All*
0.2	10 (6)	34 (17)	12 (6)	4 (2)
0.5	14 (7)	40 (20)	12 (6)	6 (3)
0.75	36 (18)	70 (35)	24 (12)	8 (4)

Example $\beta = 0.5$, 90° bend $10D$ upstream
Uncertainty $0.5\% + 0.5\% = 1.0\%$

Table 4.1 indicates some of the constraints on manufacture and installation and their effects on uncertainty of the coefficient of discharge. The table gives a few values from the standard, but ISO 5167 should be referred to for the full details of manufacturing tolerances and installation requirements. It is clear that long lengths are usually necessary to retain the uncertainty set out earlier. For a β value of 0.75 downstream of two bends in perpendicular planes a straight run of at least 78 diameters is required (allowing for both upstream and downstream lengths). An additional uncertainty is added of 0.5 per cent for lengths down to the parenthesized values.

Thus the example given of an orifice plate installed with a β value of 0.5, $10D$ downstream of a bend, results in a total uncertainty of 1.0 per cent for the meter, apart from errors due to the other parameters in equation (4.2) and the pressure measurement. To achieve a lower uncertainty it is necessary to have the orifice meter calibrated at a suitable laboratory.

Advantages of the orifice

- Well defined and documented.
- Based on long experience.
- Uncertainty calculable.
- Straightforward to install.

Disadvantages of the orifice

- Non-linear.
- High pressure loss.
- Very sensitive to installation effects.
- Very careful construction required to obtain calculable uncertainty.
- Small usable range, although increasing with smart pressure cells.
- Pulsation errors due to square law.

The summary above highlights the factors already discussed. The orifice (together with the special orifice designs, the venturi, the nozzles and the critical flow venturi described below) can be constructed from standard design guides, and these are based on much greater experience than is available for other flowmeters. Although the uncertainty is not as low as some have assumed, it is calculable with reasonable confidence. The orifice can be installed between flanges – a feature which some modern flowmeters emulate.

Some of the disadvantages of the orifice such as high pressure loss and sensitivity to installation are lessened in the venturi meter. Its range is restricted for a particular orifice size by the constraints of differential pressure measurement, although the development of smart pressure sensing is starting to remove this limitation. Although the components may appear to be straightforward to construct, the orifice meter requires great care in construction and installation. It will also require careful maintenance to ensure that the initial geometry is retained. Unless coupled with a fast-response pressure measurement system it will be particularly subject to pulsation error due to the square-root term in the flowmeter equation.

Typical applications of the orifice

- Almost any single-phase Newtonian flow (for high viscosity fluids a quadrant or conical orifice, Fig. 4.4, may be used).
- Unsuitable for abrasive fluids.
- Use with caution in two-phase fluids.

There are probably few single-phase flows for which an orifice has not been used at some time. The standards require that the fluid is physically and thermally homogeneous and of single phase. However, highly dispersed colloidal solutions such as milk are allowed by the standard.

For very viscous flows it may be appropriate to use a quadrant or a conical orifice. Abrasive fluids will, apart from other effects, wear the orifice plate and cause a change in discharge coefficient.

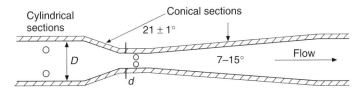

Fig. 4.3 Venturi tube – standard design

If the orifice is used in a two-phase flow it will be outside the standard requirements and should be used with caution and with suitable bleed holes to avoid liquid or gas collecting at the upstream face of the orifice plate.

The cost of manufacture and installation is increased by the need to check dimensional correctness before installation. In addition the pressure leads and the differential pressure cells must be correctly installed to allow filling, bleeding, seal liquid, etc. as required by the fluid and installation.

4.2   VENTURI METER

The standard design of venturi is shown in Fig. 4.3 with requirements on the size of cone angles. Other details may be obtained from the standard (BS EN ISO 5167). Although the venturi performance approximates to the ideal of equation (1.16), it is still necessary to introduce a discharge coefficient, C. The effect on C of the reduced losses in the flow is apparent from the value of C which for the machined version is only 0.5 per cent from the ideal Bernoulli prediction (Table 4.2).

The values of the uncertainties in C are related to the available data and the precision of the geometrical definition; to achieve a lower value it will be necessary to have the venturi meter calibrated. The pressure loss from this

Table 4.2   Venturi meter discharge coefficient

Type of convergent	*Constraints*	*C*	*Uncertainty in C (%)*
Rough cast	100 mm $\leq D \leq$ 800 mm $0.3 \leq \beta \leq 0.75$ $2 \times 10^5 \leq Re \leq 2 \times 10^6$	0.984	0.7
Machined	50 mm $\leq D \leq$ 250 mm $0.4 \leq \beta \leq 0.75$ $2 \times 10^5 \leq Re \leq 1 \times 10^6$	0.995	1.0
Rough welded sheet iron	200 mm $\leq D \leq$ 1200 mm $0.4 \leq \beta \leq 0.7$ $2 \times 10^5 \leq Re \leq 2 \times 10^6$	0.985	1.5

device is likely to lie between 5 and 20 per cent of the measured differential pressure. The throat and adjacent curvature must have a roughness less than 0.000 01 of the throat diameter. The reader should refer to the latest standards for possible relaxation of this value, for other finishes, and for the full definition of roughness.

The effect of upstream pipe fittings on the performance of the venturi, as set out in the standard, is much reduced from the orifice requirements, presumably because of the much greater flow stability in the venturi. Thus a 90° bend at 5D or more upstream should not affect the meter's performance. A gate valve fully open, an expander, and a reducer will require increasing lengths of upstream pipe separating them from the meter to retain its performance, while two bends in perpendicular planes introducing swirl will add an additional uncertainty value. It is also necessary to consider the effects of the pipework upstream of the disturbing fitting and to consult the standards on the precautions that should be taken. However, the values in the standard have been questioned and new guidelines are available from the NEL suggesting for instance, that for $\beta = 0.5$, 9D upstream length after a 90° bend may be necessary to retain predicted measurement uncertainty, and other spacings may be up by a factor of four or more (Baker 2000). The latest version of ISO 5167-2: 2002 should be consulted for recommendations on installation.

Advantages of the venturi relative to the orifice

- Lower pressure loss.
- Less affected by upstream flow distortion.

Disadvantages relative to the orifice

- Lower differential pressure for the same β value.
- Higher initial costs.
- Longer axial length of meter.

We have noted the level of the pressure loss, and this may be an important consideration in some applications. The reduced sensitivity to upstream fittings may also be important, although this is offset by the greater initial length of the venturi.

An additional advantage might be seen as the high value of discharge coefficient, which means that the performance approaches the theoretical value, and this, in turn, could suggest a more reliable uncertainty value.

The venturi has two major disadvantages: it is expensive to construct and it has a large axial length, which may be unacceptable in certain applications. For this reason various other differential pressure devices are covered in the standard, which attempt to achieve a compromise between the convenience of the orifice plate and the performance of the venturi.

Where energy conservation is an important consideration in large pumped flows, the low head loss of the venturi becomes attractive. It has been used for slurry and two-phase flows for which some data are available. New research and testing are being undertaken on its use for high pressure gas flows and wet gas measurement. However, in all these non-standard applications it should be applied with caution.

The initial cost is higher than the orifice, since it is a much larger device to manufacture.

4.3　SPECIAL ORIFICE PLATES AND FLOW NOZZLES

Apart from the orifice designs which have been discussed above, there are orifices (Fig. 4.4) with quadrant and conical inlets for viscous fluid and there are also eccentric and chordal plates which are more suited to two-phase flows, since they allow the second phase to pass, by suitably orientating the orifice. Data on these devices are available in standard documents and in some of the references in the bibliography.

Several nozzle designs are available in the standards which achieve smooth and well-controlled flow contraction, and avoid the problems of the sensitivity of the flow in the orifice's *vena contracta* to extraneous parameters. In these nozzles the downstream pressure is sensed by flange tappings or similar, but the throat is well defined by the nozzle design. They are more stable than the orifice for high temperatures and high velocities, experiencing less wear and being less likely to distort. They are particularly applicable in steam flows. To approach the low loss of the venturi without its length, a venturi nozzle is available with a divergent region which improves pressure recovery.

4.4　CRITICAL NOZZLE (SONIC NOZZLE)

Figure 4.5 shows a typical arrangement for the critical nozzle. The flow is conditioned by straightening vanes at A and B. It then passes through a

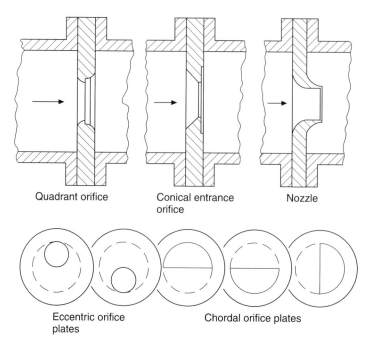

Fig. 4.4 Special orifice plates and flow nozzles

perforated plate which will create an artificial turbulence level. After this plate it enters a settling chamber. Thus, any flow disturbances from upstream will have become negligible by the time the flow enters the convergent–divergent nozzle. This is the fundamental element in the system.

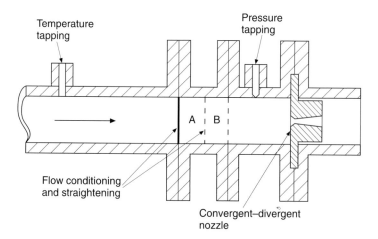

Fig. 4.5 Critical flow venturi nozzle installation

The area of flow here is much smaller than in the settling region. The key feature of the nozzle is that it is unaffected by downstream changes provided it is operating critically. The theory of compressible flow of a gas through a convergent–divergent passage shows that, for a sufficiently low back pressure, the velocity at the throat, or narrowest point in the duct, becomes sonic, i.e. the Mach number becomes unity. Because information in a gas is carried by very small pressure waves, sonic conditions at the throat prevent transmission of information from downstream. The flow is said to be choked under these conditions and changes downstream of the nozzle will not affect the flow upstream of the nozzle throat. The nozzle, for this reason, can be used as a flow control as well as a flowmeter, and it is often convenient to combine these roles.

The practical flowmeter equation is given as

$$q_{\mathrm{m}} = \frac{A_* C C_* p_{\mathrm{o}}}{\sqrt{(\mathbf{R}/\mathbf{M})T_{\mathrm{o}}}} \tag{4.7}$$

where A_* is the throat area. The discharge coefficient, C, is obtained from experimental data. C_*, the critical flow function, may be obtained from ISO 9300: 1995 or reference books (Miller 1996) for some real gases. Its value becomes C_{*i} for a perfect gas where

$$C_{*i} = \sqrt{\gamma[2/(\gamma+1)]}^{\frac{1}{2}(\gamma+1)/(\gamma-1)} \tag{4.8}$$

and γ is the ratio of the specific heats. p_{o} and T_{o} are upstream stagnation conditions. $\mathbf{R}$ is the universal gas constant and $\mathbf{M}$ is the molecular weight.

The equation defining the performance of the sonic nozzle, although empirical, is very closely related to the equation obtained from one-dimensional compressible flow theory. It is important to note that the mass flow of the gas through the nozzle is directly proportional to the pressure in the upstream chamber, provided the temperature remains constant. It is, thus, essentially a linear mass flowmeter for gases. The equation from one-dimensional flow theory is modified by the addition of a discharge coefficient, which is empirically obtained and incorporated in an international standard document. The form given in Table 4.3 is for toroidal throat venturi nozzles:

$$C = a - bRe_{\mathrm{d}}^{-n} \tag{4.9}$$

Re_{d} is the Reynolds number based on the throat diameter. Other values of C are available for cylindrical throat venturi nozzles.

Table 4.3   Parameters [equation (4.9)] for the toroidal throat critical venturi nozzle discharge coefficient, *C*

Re_d range	a	b	n	Source
10^5–10^7	0.9935	1.5250	0.5	ISO 9300: 1995
4×10^4–3×10^6	0.997 38	3.058	0.5	Arnberg *et al.* (1973)
3×10^5–10^7	0.991 03	0.0	—	Brain and Reid (1980)

ISO uncertainty ±0.5% with 95% confidence level.

Uncertainty for the critical nozzle
- The uncertainty of the instrument depends on uncertainties of all the parameters in the equation but, allowing for discharge coefficient and all other measurements, is likely to lie in the range ±1 to ±1.5 per cent.
- It should be possible to achieve an uncertainty after calibration of the order of ±0.3 per cent.

Installation
- Unaffected by upstream pipework assuming the presence of a smoothing and settling chamber.

The ISO standard lays down design details to ensure that the predicted performance is achieved. The installation of the device with gauzes and settling chamber is such that upstream fittings should have a negligible effect. It would also appear that upstream pulsation should average correctly.

Advantages of the critical nozzle
- It provides a transfer standard for gas mass flowrate.
- It is unaffected by downstream changes.
- The upstream stagnation pressure is directly proportional to the mass flow for constant stagnation temperature etc.

Disadvantages
- It will not easily cope with a wide range of flow.
- It should, therefore, be incorporated into a bank of flowmeters of different sizes.
- Its uncalibrated uncertainty may not be adequate for some gas flow measurements.
- It may need to be calibrated.

The advantages are self-evident, but the disadvantages need some further explanation. Because the flow through the nozzle can only change, assuming the temperature is constant, by varying the upstream pressure, the range of a nozzle is, in practice, rather limited. It is, therefore, customary to install a bank of nozzles of different throat sizes to allow a better range of flows. This is clearly less convenient than a flowmeter which has a wide range in one instrument. Its intrinsic accuracy may not be adequate for use as a source of calibration, so it may need to be calibrated for best performance.

Applications of the critical nozzle
- High accuracy gas flow measurement.
- Transfer standard for gas flowmeter calibration.

The applications are those most appropriate to a research and development laboratory, or a manufacturer of gas meters requiring a calibration stand. However, some people advocate its wider use. Its initial cost will be fairly high due to the precision required in making a bank of nozzles.

4.5   OTHER DIFFERENTIAL PRESSURE DEVICES

Three types of device (all proprietary makes) where pressure is measured and which depend on differential pressure principles, of varying sorts, are illustrated in Fig. 4.6. The Dall tube is similar to a nozzle or venturi, but has a recessed slot at the throat which should retain a vortex filament, and it combines low head loss with high differential pressure compared with conventional devices. Other designs such as low-loss tubes also have a similarity to the nozzle and venturi and claim values of measurement uncertainty such as ±0.5 per cent of actual flowrate. Some designs are available for mounting between flanges, like an orifice plate.

The Gilflo-B has a spring-loaded diaphragm causing a variable area orifice and resulting in a wide operating range (20:1 up to 100:1) for gas and steam, liquid natural gas, cryogenic, and other liquids. Other designs based on similar principles are also available.

Other variable-area spring-loaded meters have been used for measuring the flowrate of water, paraffin, petrol, oil, tar, distillates, etc. They can handle both high and low flows. The manufacturers' claims for factory calibration are of the order of 1–2 per cent of maximum flow.

The viscous flowmeter for gases uses very small flow passages so that the pressure drop is due to the viscous as opposed to the inertia losses and is, thus, proportional to the flowrate. It was developed to measure gas flows in internal combustion engines where there is a high level of unsteadiness.

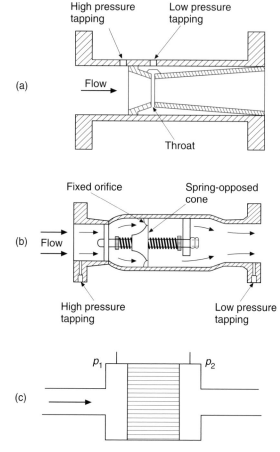

Fig. 4.6 **Approximate diagrams of other differential pressure devices. (a) Dall tube (a type associated with Kent Instruments now ABB); (b) Gilflo B spring-loaded variable area meter (a type associated with Gervase and now with Spirax-Sarco); (c) viscous/laminar flowmeter**

Advantages of the proprietary designs of flowmeter
- In certain applications proprietary devices may offer the most satisfactory option.

Disadvantages
- Little independent operational information is available and it is necessary to rely on the manufacturers' values.

The advantages will be stressed by the manufacturer. The disadvantages are that little information is publicly available and the user is, therefore, dependent on the manufacturer for information on uncertainty, repeatability, range, maintenance, application, and so on. The potential user should ensure that data for a particular application are satisfactory.

Other differential devices exist, such as wedge meters, V-cone meters, elbow meters, and swinging flap meters, and the reader will no doubt encounter even more. They make use of the momentum, or the resulting displacement due to the passing fluid, or the measured pressure loss resulting from the flow. The reader should refer to the manufacturer for advice on applications and initial cost, and if in doubt should seek expert advice.

4.6   TARGET METER (DRAG PLATE)

The drag plate meter (Fig. 4.7) is essentially an inside-out orifice plate. A plate is mounted in the pipe bore and this experiences a drag force around it. It is held against this force and the restraint exerted is used as a measure of the flowrate.

Thus, whereas in orifice meters the differential pressure across the plate is measured, here the force on the plate which is linked to the pressure drop is measured. The advantage is that this can be done directly by an electrical method, such as a strain gauge.

For an incompressible fluid where the momentum change will only be due to redistribution of the velocity profile, the force on the plate will be approximately equal to the pressure drop times area, which is related to the momentum. The force is given by

$$F = C_{D}\rho V^{2}A/2 \tag{4.10}$$

where C_{D} is the drag coefficient and A is the area of the target. This simple expression does not attempt to model the complex flow around the target meter, and implies a constancy of C_{D} which is likely to be only an approximation

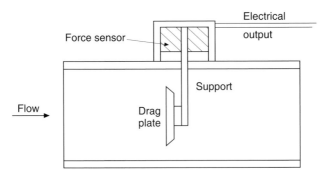

Fig. 4.7 Target meter

even for restricted ranges of Reynolds number. The target meter has been applied to two-phase flows, where it has been used with some success.

Accuracy of target meter
- If well constructed this device should achieve an uncertainty of the order of ±2 per cent of upper range value (URV).

Installation effects
- It may be assumed that the effect of upstream installation is similar to that for the orifice and it is likely to be affected by pulsation.

These comments are reasonable assumptions based on the similarity of this meter to the orifice. Because of its similarity to the orifice it may be appropriate to assume similar installation requirements, although it is said to be less susceptible to upstream disturbances, particularly misalignment. Due to the limited data on this flowmeter, it is important that information be obtained from the manufacturer on calibration, installation, application, etc.

Advantages of target meter
- Suitable for two-phase flow.
- Electrical read-out.

Disadvantages
- As for orifice.
- Shortage of operational knowledge.

Typical applications
- As for orifice where standardization is not important.
- Two-phase flows.

It is perhaps an accident of history that the orifice is widely used and not the target meter, which is essentially an inside-out orifice plate meter. Had it been otherwise we might now have seen this instrument as part of a well-specified standard backed with much data, and offering the possibility of a direct electrical output.

Because it allows gas or solids entrained in the liquid to pass, the meter has been used for two-phase flows. It needs to be used with care and understanding in this application.

4.7   VARIABLE AREA, ROTAMETER®, OR FLOAT-IN-TUBE METER

This is, commonly, a vertical, conical, glass tube with the area of the tube increasing towards the top (Fig. 4.8). Within the tube is a float which in some cases moves on an axial guide. As the flow increases the drag on the

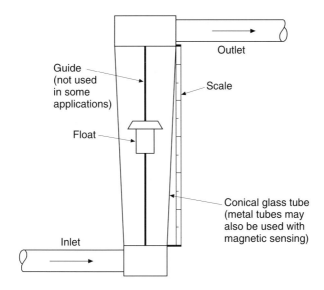

Guide
(not used
in some
applications)

Outlet

Scale

Float

Conical glass tube
(metal tubes may
also be used with
magnetic sensing)

Inlet

Fig. 4.8 Variable area meter

float increases and causes it to rise in the tube. The rise of the float to a section of the tube with a greater cross-section reduces the drag due to the annular flow around the float. The float will thus settle at a height such that the upwards drag of the flow through the annulus around the float precisely balances the gravitational force on the float. The equation governing the drag on the float is now essentially that for the target meter, except that the upward force on the float is now held constant, by the variation of the annulus area, and equal to the buoyant weight of the float.

The indication is usually by viewing the height of the float on a calibrated scale on the glass tube. However, some designs, particularly for high pressure, use a metal tube and make use of a magnetic detector or other means of turning the float position into an electrical signal. Other designs use a restoring spring rather than gravity.

The meter, with a transparent tube, is sometimes known as a Rotameter®, which is a registered trademark.

Accuracy of variable area meter
- Uncertainty after calibration of order 2.0 per cent upper range value (URV) depending on design.
- Range turndown ratio 10:1.

Installation
- It must be installed vertically.
- It may be wise to allow $5D$ of upstream straight pipe before the meter inlet.

- Sensitive to changes in viscosity.
- Sensitive to pulsation.

Advantages of variable area meter
- Visual indication of flowrate in transparent tube designs.
- Simple to install and use.

Disadvantages
- Density and viscosity sensitive.
- Low accuracy.
- Commonly without electrical read-out.
- Affected by pulsation.

This is not a highly accurate instrument and claims for small values of measurement uncertainty should be viewed with caution. It appears to be viscosity dependent, and calibration may not be retained between different fluids. In some designs the flow enters around an inlet elbow, in other designs the inlet and outlet pipes are coaxial with the conical tube. It may be prudent to include at least $5D$ of straight upstream pipe, and $2–3D$ of downstream pipe. The lift on the float is against gravity and so it is important that the tube be mounted vertically. If the flow is pulsatile the float may become very unstable. As for all momentum meters, it is density dependent.

A well-positioned variable area meter on a flow system may be seen from a distance and will provide a good indication of flow, and in this role it is most valuable. Unfortunately, many people use it as a high accuracy instrument and do not question whether it is adequate for the application proposed.

Typical applications of the variable area meter
- Liquid or gas flowmeter where a high accuracy instrument is not required.
- Visual indicator of flow and flowrate.

4.8 MOMENTUM-SENSING FLOWMETERS – GENERAL COMMENTS

Differential pressure devices are often supposed to be of high accuracy and simple construction. This is not necessarily so. High accuracy without calibration (0.5 per cent uncertainty at best for the coefficient of discharge for the orifice, plus pressure measurement and other uncertainties) is only achieved with great care and precise manufacture. It is only maintained by regular inspection to ensure that the installation continues to meet the requirements of the standard.

Because differential pressure devices depend on the momentum change of the flow, which is proportional to the square of the velocity, there is a 'square-root' error introduced when the flow is unsteady or pulsatile. The inertia error is usually negligible in comparison with this one. However, the read-out device may also be sensitive to unsteady flows. Pressure (impulse) tubes are notorious for introducing errors in the measurement of fluctuating pressures.

The situation with proprietary instruments is rather different. There is no publicly available mass of data to support the claims made for the instruments. It is, therefore, important that the manufacturer supplies evidence of performance, so that the user may have confidence that the meter will come up to requirements. It is also important that the user knows how often the instrument should be inspected, overhauled, and recalibrated. This information must be sought from the manufacturer.

Another momentum-sensing device is discussed in Chapter 7. This is the averaging pitot differential pressure insertion probe or spool piece meter.

CHAPTER 5

Volumetric Flowmeters

Positive displacement meters
- For liquids.
- For gases.

Turbine meters
- For liquids.
- For gases.

Oscillatory meters
- Vortex shedding meters.
- Swirl meters.
- Fluidic meters.

Electromagnetic meters

Ultrasonic meters
- Transit time.
- Doppler.
- Correlation.

5.1   POSITIVE DISPLACEMENT METERS

Description of operation

The measured volume and stop-watch still offer one of the most accurate methods of flow measurement, and this is the principle of the positive displacement meter. In each meter of this family, the rotation carries fluid through the meter in, essentially, self-contained compartments. These closed compartments are formed either by means of sliding vanes which seal against the meter surface of the chamber, or by means of specially shaped rotors which keep contact as they rotate and cause a certain metered volume to be passed through the meter from inlet to outlet. An alternative is to use pistons in cylinders, or some other form of linear mechanism. The volume passed is, thus, proportional to the rotation and the volumetric flowrate, q_v, is proportional to the rotational speed.

5.1.1   Positive displacement meters for liquids

The rotation of the meter allows well-defined volumes of liquid to pass through the meter in, essentially, closed compartments as described above.

Three designs of *multi-rotor meters* are shown in Fig. 5.1; in these meters the rotors seal against each other. In some designs both rotors transmit fluid while in others one rotor transmits net fluid while the other provides a seal to return the rotor blades.

The *oval gear meter,* Fig. 5.2, is similar to the multi-rotor meters except that the seal between the rotors is enhanced by gear teeth on each rotor which also ensure correct relative rotation. As well as the oval gear version, there are also circular gear meters where the liquid is carried through by the teeth on the outer part of the wheel while the teeth in the centre mesh together to reduce leakage.

The *nutating disc meter* and the *oscillating circular piston meter* (also known as the *rotary piston meter)* are shown in Figs 5.3a and b. The nutating disc is constrained by a central bearing and by the transmission to nutate. It is prevented from rotating by a partition which separates the inlet and

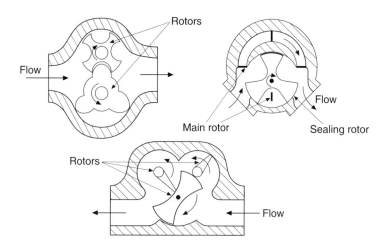

Fig. 5.1 Three designs of multi-rotor meters for liquids

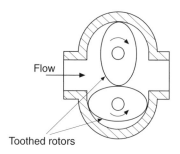

Fig. 5.2 Oval gear meter for liquids

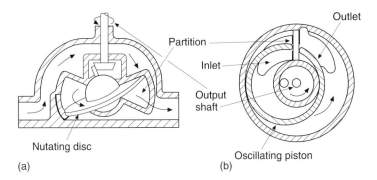

Fig. 5.3 (a) Nutating disc meter for liquids; (b) oscillating circular piston meter for
liquids (rotary piston meter)

outlet streams. The oscillating circular piston is similar in that the rotation is
constrained by a partition causing a rocking motion, which in turn causes the
shaft, on which it is eccentrically mounted, to rotate.

Sliding vane meters appear to be among the most accurate. In these meters
the vanes move radially out to form compartments and retract to release fluid
and then move back to the inlet side. The main difference between the designs
shown in Figs 5.4a and b is that one operates on an internal cam while in the
other the vanes are controlled by a contour within the measuring volume.

The *helical rotor meter*, Fig. 5.5, again traps liquid between the rotating
members which are of complex shape. Its operation is hard to visualize
without the benefit of a working model.

The *reciprocating piston meter*, Fig. 5.6, uses a system of open and closed
ports to meter liquid via the volumes of each cylinder. In the example shown the
adjacent cylinders provide the valve operation by the position of the pistons.

For very high accuracy designs, leakage can be a problem where low
rotational speeds allow the leakage to be a substantial fraction of the flow.
Where a small value of measurement uncertainty across the flow range is

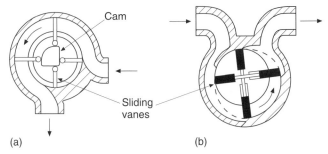

Fig. 5.4 Sliding vane meters for liquids: (a) operating on an internal cam; (b) operating
on a contour within the measurement chamber

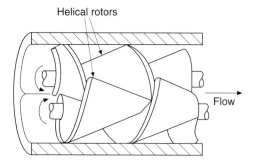

Fig. 5.5 Helical rotor meter for liquids

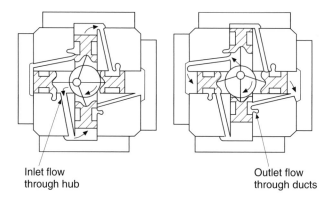

Inlet flow
through hub

Outlet flow
through ducts

Fig. 5.6 Reciprocating piston meter for liquids

essential, the meter is subjected to *slippage* tests to measure the size of the leakage which occurs at very low speeds of rotation.

Accuracy of positive displacement meters for liquids

Typical uncertainty after calibration will be of the order of:

- for nutating disc meter   ±2.0 per cent of rate;
- for oval gear meter   ±0.5 per cent of rate;
- for sliding vane meter   ±0.15 per cent of rate.

Range turndown ratio is likely to be of the order of:

- for high accuracy meters   about 10:1 or greater;
- for water meters   100:1 or more.

For high accuracy meters uncertainty is probably limited by the calibration rig. To retain high performance, a regular schedule of calibration checks every 6 months should be instituted, and the period should be halved if the uncertainty values are not retained by the end of the period.

Installation effects/requirements for positive displacement meters for liquids

Installation effects/requirements for high accuracy meters are likely to be as follows:

- negligible effect from upstream flow profile;
- flows should be filtered to remove particles which can cause wear and block the meter;
- gas bubbles can cause errors;
- temperature compensation may be necessary;
- double case meters should be unaffected by pressure changes.

Of all the common meters the positive displacement meter is one of the few virtually unaffected by flow profile. However, it is a high accuracy instrument and not only must the fluid be particle free, but gas coming out of solution will severely affect readings, and temperature change will cause expansion of the chambers which, without compensation, may cause about 0.1 per cent change in 20 °C. Liquid density change is at least ten times greater than this. Some meters are double skinned so that the measuring volume is unaffected by pressure differential. Where this is not the case 30 bar change will result in a change of the order of 0.1 per cent in the measuring volume. Changes in these parameters will also cause changes in leakage past the rotors.

Advantages of positive displacement meters for liquids

- High accuracy at one end of the range with application to transfer standard use.
- High turndown ratio for utility flows at the other end of the range.
- Pulse output (mechanical or electrical) proportional to flowrate within uncertainty limits.

Disadvantages

- High accuracy versions can be bulky.
- Can cause total line blockage.
- Can be damaged by sudden flow change.
- May corrode with water – when not in use an inhibiting agent may be used for protection.
- May create flow pulsation.

The high accuracy design offers a meter that can be used to calibrate other devices, since it will retain its own calibration for long periods and during transportation to other sites.

The best of these devices are extremely accurate and need to be treated and maintained with great care. When not in use they should be protected by a suitable inhibiting agent. They should not be subject to violent transients and pulsating flows should be avoided. At very low flows their performance is less good and *slip* occurs, when more liquid passes than is registered. It is, therefore, important to restrict their operation at very low flows, and to test the extent of slip when checking the meter's performance.

For the dispensing of hydrocarbon liquids the positive displacement meter has to operate over a wide flow range. To avoid sudden start and stop of the flows which could cause violent pressure waves and damage the meter, the increase and decrease is achieved with a controlled acceleration and deceleration of the flow.

The meter is large compared with other devices for the same capacity. It may also block the line if the vanes become jammed by particulate matter.

Typical applications of positive displacement meters for liquids

- Metering hydrocarbons (multi-rotor, sliding vane).
- Calibration and transfer standard duties (sliding vane).
- Oil hydraulics (gear).
- Chemical dosing (reciprocating piston).
- Milk metering (oscillating circular piston).
- Water metering (oscillating circular piston, nutating disc).

It is particularly valuable for flows of high value liquids where a small value of uncertainty with prior calibration is required. It is frequently used as a transfer standard and is installed in calibration rigs to provide a source of calibration for a flowmeter manufacturer. The rotary piston, and possibly other designs, are available for hygienic applications. Lower cost designs are used to measure the low flows of various liquids.

Existing designs are unsuitable for slurries and multiphase flows, apart from one or two commercial designs specifically aimed at the measurement of multiphase flows from oil wells.

5.1.2 Positive displacement meters for gases

Description of operation

As for the liquid version the rotation of the meter allows well-defined volumes of gas to pass through the meter in, essentially, closed compartments. The volumes are contained to ensure that a certain metered volume is passed through the meter. Transmission of rotary movement may be by means of a magnetic coupling to avoid sealing problems. Some meters have an option of pulse output enabling remote reading of flowrate and gas volume.

The *diaphragm meter* is commonly used for domestic gas metering in the United Kingdom, Fig. 5.7. It is essentially a piston meter in which compartments A, B, C, and D are the cylinders and the concertina provides a common piston for A and B and similarly for C and D.

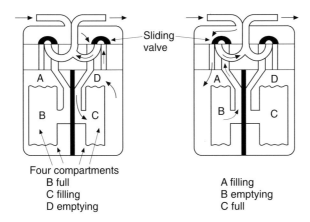

Sliding valve

Four compartments
  B full
  C filling
  D emptying

A filling
B emptying
C full

Fig. 5.7 Diaphragm meter for gases

The *wet gas meter* uses a water bath as the gas seal to create closed compartments for the transfer of the gas, Fig. 5.8. It is an early design which was developed for metering gas into a customer's plant or buildings, but has proved to have a high accuracy if used with care, and is still used as a transfer standard by some calibration laboratories.

In the *rotary displacement meter*, Fig. 5.9, two rotors mesh and rotate within an oval body contour so that leakage is at a minimum between the rotors,

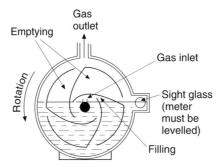

Emptying

Gas outlet

Gas inlet

Rotation

Sight glass
(meter must be levelled)

Filling

Fig. 5.8 Wet gas meter

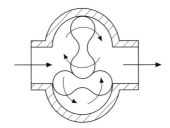

Fig. 5.9 Rotary positive displacement meter for gases

and gas is transferred at the outer point of the oval. This meter is, sometimes, referred to as a Roots® meter, a registered trademark. It is increasingly widely used and reputed to have a very high performance and turndown ratio.

The *CVM meter* is similar in concept to the multi-rotor liquid meter with sealing rotor.

Accuracy of positive displacement meters for gases

Probable uncertainty after calibration is of the order of:

- for rotary displacement meter   ±0.5–2 per cent of rate;
- for wet gas meter   ±0.35 per cent of rate;
- for diaphragm meter   ±1–3 per cent of rate.

Range turndown ratio is likely to be of the order of:

- for the rotary displacement meter   10:1 possibly up to 80:1;
- for the wet gas meter   10:1;
- for the diaphragm meter   160:1 or more.

Installation effects/requirements

Installation effects/requirements for these meters are likely to be as follows:

- negligible from upstream pipework;
- pulsation may affect performance and meters may cause pulsating flow effects;
- diaphragm meters have restricted temperature range near ambient reflecting their application to domestic gas metering.

The rotary displacement meter may be capable of achieving better than the uncertainty values in the grey box above, over a range of 10:1 or considerably higher. The author has heard reports of good performance over a 80:1 turndown. The wet gas meter is a high accuracy meter, but it tends to be rather bulky and is mainly used for calibration purposes today. The diaphragm meter is primarily used as a domestic gas meter and has an uncertainty of about 2 per cent on a 160:1 range (or greater) to accommodate the range of domestic gas flows, at one extreme for pilot flames and at the other extreme for boilers etc. It retains its performance over long periods and operates close to seasonal ambient conditions. The CVM meter is claimed to be capable of better than 1 per cent uncertainty. Most of these meters are available with a mechanical output although, increasingly, there is likely to be an electrical option.

Advantages of positive displacement meters for gases

- High accuracy (rotary and wet gas).
- Large turndown (diaphragm).
- Transfer standard use (wet gas).

Disadvantages of positive displacement meters for gases

- May cause line blockage.
- May cause pulsation.

Typical applications

- Metering gas in national networks (rotary).
- Domestic metering (diaphragm).
- Transfer standard (wet gas).

5.2 TURBINE METERS

Relationship between rotation and flowrate

The relationship for this family of meters approximates closely to a linear relationship between volumetric flowrate, q_v, and the shaft rotational speed of the meter. However, this relationship will only be valid within certain uncertainty limits over a range defined by upper and lower range limits.

5.2.1 Turbine meters for liquids

Description of operation

The propeller and screw have a long and distinguished history in flow measurement. The aim is to ensure that the blades cut through the fluid with as little disturbance as possible, so that they 'cut' a helix in the fluid, the pitch of which will relate flowrate to revolutions per second. Provided the bearing drag is small and the blades are well designed, revolutions of the turbine wheel will give a good measure of the flow past the wheel. The revolutions may be measured by various means, but for highest accuracy a low drag method is needed.

The liquid flowmeter ranges from instruments of very high accuracy, such as that shown in Fig. 5.10, to extremely robust lower accuracy water meters, such as the Woltmann-type meter in Fig. 5.11. Although the basic concept is similar the application is rather different. The main components of the former are shown in Fig. 5.10. Note that the hangers which are necessary to position the bearings centrally in the tube, also do service as a flow straightener. The rotation is usually sensed by magnetic pick-ups of various types or, in very low-drag designs, radio-frequency pick-ups. These give an electrical signal in the form of a pulse train, which can be transmitted and counted. It is important to be certain of the integrity of the signal to ensure that spurious noise does not result in additional spikes and an error in the measurement. In

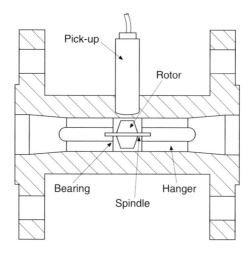

Fig. 5.10 High accuracy liquid turbine meter

the water meter the rotation is usually by means of a mechanical read-out connecting the wheel with a worm gear drive, see Fig. 5.11. Because the transmission is mechanical, it is difficult to incorporate a calibration adjustment, and so a flow deflector, or rudder, is positioned upstream of the wheel, so that the angle can be adjusted to give a rather crude calibration adjustment.

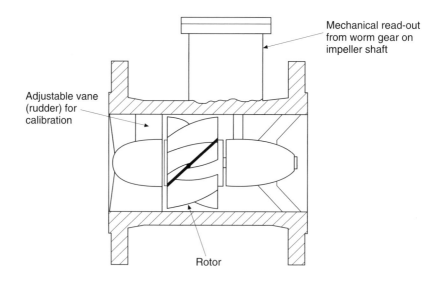

Fig. 5.11 Woltmann-type mechanical water meter

Some liquid flowmeters use flat-section blades, Fig. 5.12a, and ideally these will cut smoothly through the flow in a perfect helix. An aerofoil at zero incidence causes an obstruction around which the flow accelerates. With increasing incidence the velocity on the top (or suction) surface increases with a consequent (see Bernoulli's equation) drop in pressure. On the lower (or pressure) surface the velocity is reduced and the pressure increases. Hence, a lift force is created across the aerofoil. The drag will be due to friction and pressure loss. For low angles of incidence the drag is small; as the incidence increases the drag increases. Figure 5.12b shows how the blade creates increasing lift with incidence angle. When the incidence of flow on to the blade is too great the flow separates from the blade and stall occurs (Fig. 5.12c).

Relation between rotation and flow
Figure 5.13a gives a triangle for ideal flow angle on to the blade which relates the axial flow velocity, V_{ax}, the blade velocity, V_b, the relative velocity

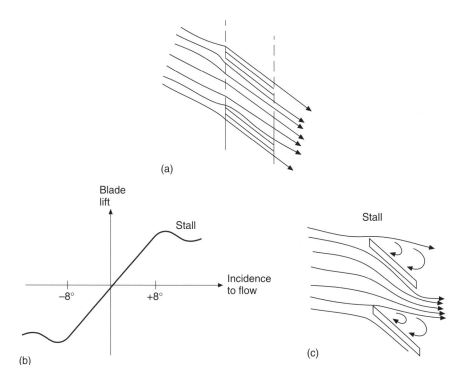

Fig. 5.12 Turbine blade behaviour. (a) Small incidence angle streamline flow (relative to blades); (b) blade lift for a flat plate at a few degrees of incidence to relative approaching flow; (c) large incidence causes flow to separate with high drag (stall)

of the fluid passing over the blade, V_{rel}, and the blade angle, β, and from this we obtain the equation

$$V_{ax} = V_b/\tan \beta \qquad (5.1)$$

and

$$f = N \tan \beta \, V_{ax}/2\pi r \qquad (5.2)$$

where f is the frequency of blade passing, N is the number of blades, and r is the radius of the blade section under consideration.

Unfortunately the theory is not as simple as this, as the blades do not cut the fluid perfectly. If the blade is of constant angle from hub to tip, then it is likely that the flow will not pass over the blades at all radii correctly. The angle of flow on to the blade will vary and may lead to separation of the flow. Thus, the angle of the helix, and thus the value of β, may vary with radius from hub (r_h) to tip (r_t) of the blades to accommodate the profile of flow across the pipe, and to present the correct angle to the inlet flow. Figure 5.13b indicates the possible angle variation for a design in which the blades are twisted to allow for this. For a uniform

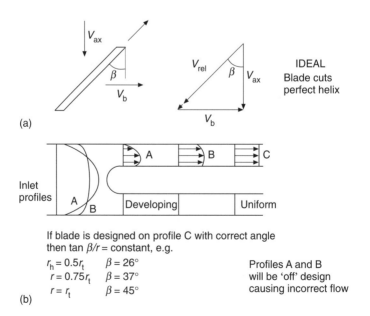

Fig. 5.13 Blade angles and flow triangles. (a) Flow triangle; (b) blade angle variation across annulus

profile, with no variation in axial velocity across the annulus, the blade angle changes from 26 to 45° from hub to tip for a radius ratio of 1:2. If, however, profiles A or B existed, then flow would not meet the blades at the correct angle.

It thus becomes apparent why the flowmeter will be susceptible to incorrect installation, since this will cause a flow profile which results in the wrong incidence angles at the blade for some parts of the annulus. It also suggests why a turbine wheel that is optimized for flow profile will give a better performance than one with constant angle blades. The correct or incorrect flow over the blades will, therefore, have an important effect on the performance of the meter and will account for the effects of upstream flow disturbance, viscosity effects, and Reynolds number effects. It is likely that a meter with twisted blades of the correct angle will have less susceptibility to viscosity change than one with flat untwisted blades.

A well-designed turbine meter should, therefore, operate so that the incidence at all radii is small, and at no time should any blade sections approach a stalled condition. In order to achieve this it will also be necessary to design the meter with bearings which minimize the drag, and with a sensing pick-up with negligible drag. Such a meter may then achieve a linear range of the order of 10:1 with a small value of measurement uncertainty.

Bearings for high accuracy turbine meters are:

- ball bearings (process liquid lubrication) for clean liquids such as cryogenic and petroleum liquids;
- self-balancing journal bearings (material: tungsten carbide, sapphire, PTFE, etc.) for clean liquids such as water and lubricating liquids;
- journal bearings with thrust ball for other liquids;
- 'bearingless' for aggressive and non-lubricating liquids.

The bearings are a critical part of the meter and must be of a suitable design and material for the application. Tungsten carbide is probably the most common bearing material, but titanium carbide, stellite, ceramics, etc. are also used. Bearing wear will affect the meter calibration.

Turbine meter bearings may be damaged if subject to fluctuating flows for extended periods. The bearings may also be designed for particular fluids, and may be damaged if used with other fluids. Special designs are available in which the rotor is 'bearingless' and, in one design, had two blade rows which were supported by the flow.

The drag is also influenced by the tip clearance between the ends of the blades and the turbine casing. A compromise must be achieved between too small a clearance, leading to high drag, and too large a clearance, allowing some of the flow to pass the turbine without being 'metered'.

In practice the turbine meter closely approximates to a linear meter for high flowrates. Two factors are commonly used to describe the turbine operation:

- The K factor relates flowrate to the frequency of pulses and is given by

$$K = \frac{\text{pulses}}{\text{unit volume}} \qquad (5.3)$$

- The meter factor is usually defined as

$$\text{Meter factor} = \frac{\text{true volume}}{\text{indicated volume}} \qquad (5.4)$$

The reader should keep a wary eye for other definitions of meter factor such as the reciprocal of the K factor.

Accuracy
Figure 5.14 shows a typical industrial performance curve for a liquid flowmeter. Note the three main ranges. The low end of the characteristic can be seen to fall off rapidly, and flow measurement or calibration in this region should be attempted with great caution.

Typical ranges and measurement uncertainties for turbine meters for liquids

- Highest precision range 5:1   typically ±0.25 per cent of rate.
- Normal operating range 10:1   typically ±0.5 per cent of rate.
- Extended range for short period   typically ±0.5 per cent of rate.
  operation 15:1

The turbine meter can be calibrated to an accuracy of the order of 0.25 per cent over a flow range of the order of 5:1 and it, therefore, provides an attractive method of fiscal metering, where frequent recalibration can be built into the system using, for instance, a pipe prover. It is also used as a transfer standard, and in this application it is best calibrated with upstream pipework and flow straightener to minimize the effects of upstream disturbances. However, recalibration is necessary due to bearing wear in particular. The water meter is a more robust, lower accuracy instrument of much greater range.

With modern signal processing, the linearity becomes less important as the calibration curve will be stored and flow signals corrected from it. However, it is important not to allow this to hide the greater uncertainty at the extremes of the range.

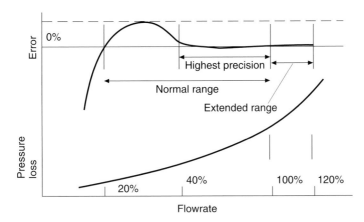

Fig. 5.14 Typical performance characteristics for liquid turbine meter

---

Likely installation effects/requirements for high accuracy turbine meters for liquids

- Sensitive to swirl since this will have relative rotation to the blades.
- Sensitive to profile distortion due to blade stall and flow through tip gap.
- Sensitive to viscosity.
- Sensitive to particles in flow affecting bearings.
- Sensitive to cavitation.
- To maintain a meter's high accuracy it may be necessary to install a flow straightener 10D upstream.

In terms of Table 3.2 sensitivity is likely to be medium/high.

The turbine meter is sensitive to swirl, particularly since swirl creates a relative rotating environment, but the meter is also sensitive to other upstream disturbances, and to viscosity variation.

The available data from turbine meters in two-phase flows suggest that they are subject to large errors (Baker 2000). The reason for this is unclear, and so no operating guidelines can be suggested. Solid particles may damage the flowmeter and steam may cause erosion. Flowmeters for hygienic applications can also be damaged by overspeeding when steam purging is used.

Advantages of turbine meters for liquids

- Very high accuracy instruments used for fiscal measurements and custody transfer.
- Linear over 5:1 range or more. A meter with twin rotors claiming to have a very low drag bearing and pick-up arrangement has been available.
- Robust mechanical water meters are also available.

Disadvantages of turbine meters for liquids

- Use with upstream straightener and filter for best results.
- Affected by viscosity (avoid use with liquids of viscosity above 25 cSt for highest performance).
- May over-read in pulsating flows.
- Recalibrate regularly due to bearing wear.
- Can be damaged by overspeeding.

Typical applications

- Single-phase liquids without extreme flowrates, either high or low.
- High accuracy meter for liquid flows and custody transfer of liquids.
- Water meters widely used for national networks.
- Pelton wheel designs for very low flows.

The flowmeter is widely used to monitor liquid hydrocarbon flows in North Sea oil pipelines, and is usually installed with a prover to allow regular recalibration. In high accuracy instruments two pick-ups are common to provide a check value. The flowmeter needs careful installation and maintenance.

Designs using a blade row which is like a pelton wheel (blades more like buckets) allow very low flows to be measured. Several designs position the pelton wheel or propeller in a by-pass line across an obstruction such as an orifice plate. The flow in the main line will then be proportional to the flow in the by-pass, but the meter will need to be calibrated.

As discussed under the section on the gas turbine meter, there is a possibility that the meter will over-read in unsteady or pulsating flows.

5.2.2 Turbine meters for gases

Description of operation

The operation of the gas meter is, essentially, the same as that of the liquid meter. The differences are due to the relative size of the hub and the blades. In these high accuracy instruments, the diameter of the hub is large and the flow passage relatively small. This ensures that the flow is straightened by the hangers, and any swirl angle is reduced by the increase in the radius. The flowrate is larger due to the small passage, and the large hub results in a larger torque on the wheel. The transmission is often by a mechanical gear train, but there will usually be an electrical pulse option also. Figure 5.15 is a diagram of such a meter. Calibration adjustment may be possible with a deflection vane upstream of the turbine wheel, using the gear train, or through the electrical pulse output. Rotary movement may be transmitted via a magnetic coupling.

Many designs of gas meter use ball bearings with lubrication.

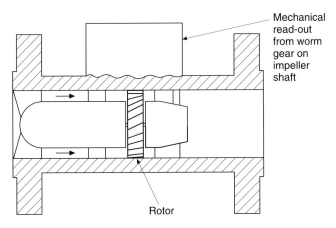

Mechanical read-out from worm gear on impeller shaft

Rotor

Fig. 5.15 Gas turbine meter

Simplest equation
The equation is, essentially, the same as equation (5.2) for liquid meters.

Accuracy for turbine meters for gases
- Measurement uncertainty after calibration for gas meters may be within the range ±0.5–1.5 per cent of rate.
- Calibration should be checked within 6 months (due to bearing wear).
- Linearity for high accuracy gas meters should be better than ± 1.0 per cent of rate.

Installation effects/requirements
For high accuracy meters these are likely to be as follows:

- some sensitivity to swirl and profile distortion but possibly less than for liquid meters;
- unsteady and pulsating flows are likely to cause the meter to over-read;
- viscosity variation is less important than for liquid meters;
- density variation will change the operating characteristic.

A gas meter may be available with a second rotor in close proximity to the main rotor, which senses the rotation of the flow leaving the main rotor, and uses this to check that installation, performance, etc. have not changed since calibration and initial installation.

The reason for the turbine meter's sensitivity to varying flows appears to be that the increasing flow creates higher incidence angles on the turbine blades and the turbine wheel accelerates fast. However, when the flow decreases the blades, presumably, stall with low lift and hence low deceleration. The effect is to give an overestimation of the total flow. The same occurs with pulsating flow, since, again, the turbine flowmeter cannot follow the fluctuating flow and reads high.

Advantages of turbine meters for gases

- Very high accuracy instruments used for fiscal measurement and custody transfer.
- Mechanical and pulse output proportional to flow.

Disadvantages

- Gas flowmeter over-reads in pulsating flow.
- Sensitive to gas density.

Typical applications

- High accuracy instrument for control of gas networks and for custody transfer of gas.

In one gas installation with a varying flow with a cycle of about 100 s – during which the flow dropped to almost zero, rapidly increased to a high value, rapidly dropped to an intermediate value, and then repeated the cycle – the author estimated an over-reading of a few per cent.

5.3   OSCILLATORY METERS

Three main oscillating fluid meters are available: the vortex meter, the swirl meter, and the fluidic flowmeter. The first may be the most widely used, but the last could become very widely used if its promise for utility flows is realized. The swirl meter has a niche for a number of important applications.

Relationship between frequency and flowrate
These are essentially linear meters for which the volumetric flowrate, q_v, is proportional to the frequency signal which results from the fluid oscillation. However, this relationship will only be valid within certain uncertainty limits over a range defined by upper and lower range limits.

5.3.1   Vortex shedding meters

Description of operation
The vortex flowmeter consists of a bluff body within a circular pipe, which sheds vortices alternately from each side, in the same way as the Karman vortex street behind a circular cylinder. The width of the body is about 25–30 per cent of the pipe diameter, and the familiar vortex shedding behind a cylinder is greatly stabilized by the short body with, effectively, end plates formed by the pipe wall. In addition the body has very well-defined edges from which the vorticity is shed, unlike the circular cylinder. A cut-away

view of a typical meter is shown in Fig. 5.16a. Typical bluff bodies are shown in Fig. 5.16b; the most common shapes approximate to a rectangle or a triangle, although there is some suggestion that dual bluff bodies may have advantages.

Sensors depend on various physical parameters, and some will be more suited to particular fluids.

Examples of sensors for vortex meters
- Thermistor: either in upstream facing or in transverse flow duct.
- Pressure: sensed by diaphragm or by moving tail.
- Capacitance: located in a fin behind the bluff body and sensing very small deflections.
- Shuttle ball: movement sensed magnetically.
- Strain gauge: on deflector beam.
- Ultrasonic: modulated by vortices.
- Optical fibre: in bluff body.

Some sensors are mounted in a second body downstream of the shedding body. The instrument appears to have a similar characteristic for liquids and gases. Because of the wide range possible, the pressure fluctuations may have a range of 100:1 or more, and the change from liquid to gas will extend this. It must, therefore, be the objective of manufacturers to produce a design which, with a single sensor, covers this range. Ultrasonics or, for water-based liquids, electromagnetic sensing, may be the best possibilities.

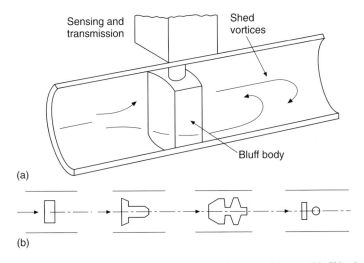

(a)

(b)

Fig. 5.16 **Vortex flowmeter. (a) Cut-away diagram of meter; (b) some bluff body shapes**

Simplest equation

Figure 5.17 shows a diagram of the shedding mechanism. The Strouhal number, *St*, is an important parameter that relates the frequency, *f*, with which vortices are shed for a particular body design and shape, with flowrate. For such a meter the Strouhal number is given by

$$St = fw/\overline{V} \tag{5.5}$$

where *w* is the width of the bluff body and $\overline{V}$ is the mean velocity in the pipe. A typical ratio of *w/D* (where *D* is the pipe diameter) is 0.3, and values of *St* lie, typically, in the range 0.24–0.27 for vortex meters.

The *K* factor (pulses/unit volume) is a measure of the performance used for this type of meter also [see equations (5.3) and (5.4)].

Accuracy and installation effects/requirements for vortex meters
Claims for accuracy are varied, and the value suggested will be disputed as both too high and too low:
- probable uncertainty ±0.5 to ±1.5 per cent of rate;
- linearity probably of the order of ±1.0 per cent of rate;
- turndown ratio of 10:1 or greater;
- minimum Reynolds number (for stable shedding) 10 000–30 000.

Installation effects/requirements for high accuracy meters are likely to be as follows:
- medium to high (Table 3.2);
- no protuberances upstream due to gaskets, weld beads, thermometer pockets, etc. and no pipe diameter changes.

Claims for measurement uncertainty are varied. Claims have been made for very large operating ranges, but 15:1 is probably achievable, and may be exceeded in some designs.

The meter appears to be very sensitive to upstream disturbance, presumably because any fitting upstream will change the vorticity within the flow and this will affect the vortex shedding. Some manufacturers even emphasize

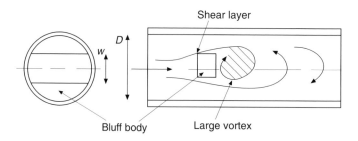

Fig. 5.17 Vortex shedding mechanism

the importance of a smooth pipe upstream, free of protuberances and diameter changes. Great care should be taken if the calibration of the meter is to be retained.

The range of operation of these meters is limited by the strength and stability of the shedding for low Reynolds numbers. A minimum value for Reynolds number is usually in the range 10 000–30 000. There may be problems with compressibility for gases and cavitation for liquids at the upper end of the range. Their size range is also limited since, as the size of the bluff body increases, so the shedding frequency falls until it is too low to obtain a reasonable speed of response.

Advantages of vortex meters
- Linear frequency output.
- Low sensitivity to temperature, viscosity (for Re > 20 000) and density.
- Suitable for liquid and gas.

Disadvantages
- Good installation essential.
- Not suitable for multiphase flows.
- Flow pulsations may cause errors.
- Cavitation and compressibility errors at extreme flowrates.
- Low flowrates limited by shedding stability and high flowrates in large pipes limited by low frequency

Typical applications

Similar to those for orifice plates but different designs need to be checked with the manufacturer:
- liquid flows and gas flows in pipes of 15–300 mm diameter for water, milk, hydrocarbons, air, other gases, etc.;
- superheated steam;
- cryogenic applications.

This meter is comparable with the orifice plate meter. It has an area ratio equivalent to a β value of about 0.8 and an equivalent pressure drop of a similar order. It lacks the mass of data that the orifice plate has acquired through many years of testing, but it has the great advantages of linearity, electrical frequency output, and wider range. The range, although limited by vortex shedding stability, is several times greater than that for the orifice. It appears to be relatively insensitive to fluid properties and some tests have indicated calibrations on gas and liquid to be virtually the same.

The flowmeter is also sensitive to spurious frequencies such as flow pulsation. A second phase, even if it leaves the vortex shedding unaffected, may upset the sensing element.

Applications are similar to those for orifice plates and, in a similar way, will be limited by experience. However, each design of vortex meter will have a different sensitivity to particular liquids and gases, and no general rules will be usable.

The device has begun to find a niche, in particular, for steam measurement. However, its sensing systems may limit full exploitation of its range, and it will be in competition with other alternatives to the orifice plate.

5.3.2   Swirl meters

Description of operation

Figure 5.18 shows a diagram of the swirl meter. As the flow enters the meter, inlet guide vanes cause it to swirl. As the flow moves through the contraction, the angular momentum created by the guide vanes is largely conserved, and so the angular velocity will increase. The vortex filament develops into a helical vortex, as shown in the diagram, which moves to the outside of the tube. The frequency with which the helix passes the sensor provides a measure of the flowrate. The swirling motion is removed by vanes at exit. The sensing is usually done by means of two diametrically opposed pressure sensing ports.

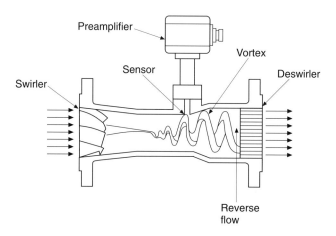

Fig. 5.18 Swirl flowmeter (from Baker 2000 reproduced by kind permission of Cambridge University Press after Bailey-Fischer & Porter with their permission, now ABB)

Accuracy of swirl meters

- Measurement uncertainty better than 1 per cent of rate is claimed by the manufacturer:
  - for the upper 80 per cent of the range for small sizes;
  - for the upper 90 per cent of the range for sizes above 32 mm.

Installation effects/requirements

- Installation effects/requirements claimed by the manufacturer for the meter are likely to be low (Table 3.2):
  - $3D$ upstream of the inlet flange;
  - $1D$ downstream for bends and expansions;
  - contractions downstream not permitted.
- Maximum viscosity 70 mPa s (70 cP).

Advantages of swirl meters

Manufacturer's claims are:

- reliability with minimum maintenance;
- in certain circumstances it may be self-cleaning.

Disadvantages

- Pressure drop may be large.
- Cavitation should be avoided.

Typical applications

- Wet and dirty gases or liquids.
- Water, sludge water, condensate, acids, solvents, petrochemicals.
- Air, CO_2, natural gas, ethylene, steam, etc.

5.3.3   Fluidic meters

Description of operation

This device is of particular interest for the utilities as it appears to be capable of a large turndown ratio. Referring to Fig. 5.19, the jet from the inlet duct oscillates between directions A and B due to the feedback via channels A′ and B′. Its frequency of oscillation is proportional to the volumetric flowrate. It is capable of operating at low flowrates. One design for domestic water metering uses permanent magnets to induce a varying voltage in the flow from which the frequency is obtained. This allows a very large turndown ratio for the meter. Other sensing methods are more restricted, and may limit turndown.

The relationship for this meter approximates closely to a linear relationship, as for the vortex meter, between frequency and flowrate, q_v.

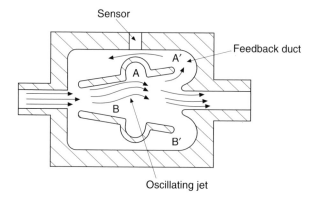

Fig. 5.19 Fluidic flowmeter

Accuracy of fluidic meters

- Uncertainty after calibration is probably achievable within ±2 per cent of rate.
- Turndown ratio may be as high as 250:1 or more.

Installation effects/requirements

- Installation effects for this meter are likely to be low to negligible, since:
  - the flow will pass through a small-diameter tube;
  - it may have a built-in flow conditioner in the path.

Advantages of fluidic meters

- Large range from low flows.

Disadvantages

- Adequate sensors to cover the range for non-conducting liquid or gas applications.

Typical applications

- Utility flows in the water and gas industry.

5.4 ELECTROMAGNETIC METERS

Description of operation

When a fluid flows through a non-magnetic tube in a transverse magnetic field (Fig. 5.20), voltages and currents are generated in the fluid due to the motion. The inner surface of the flow tube is, therefore, covered with an insulating liner to avoid shorting out the small voltages. If the voltage is measured between two electrodes in the pipe wall, it will provide an indication of the volumetric flowrate in the pipe. The correct design of the pipe and

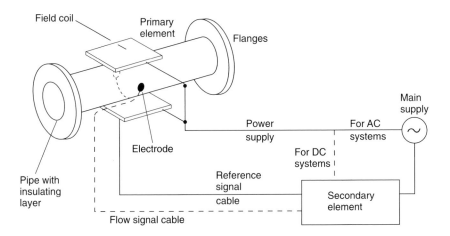

Fig. 5.20 Electromagnetic flowmeter primary and secondary elements (from Baker 2000 reproduced by kind permission of Cambridge University Press)

magnetic coils is essential to achieve a flowmeter which is little affected by upstream disturbance.

Relationship between signal and flowrate

The relationship between the voltage generated, ΔU, and the mean velocity, $\bar{V}$, for this family of meters approximates closely to a linear relationship in which

$$\Delta U = SBD\bar{V} \tag{5.6}$$

where S is the sensitivity and is dependent on magnetic field shape and flow profile, B is the magnetic flux density, and D is the pipe diameter. However, this relationship will only be precisely valid if:

(a) the magnetic field is uniform;
(b) the flow profile is axisymmetric.

If this is the case then $S = 1$ and the meter approaches the ideal, in that it gives a signal proportional to the mean velocity or the volumetric flowrate.

One additional aspect of the design which it is important to be aware of is the concept of the weight function. This was first proposed by Shercliff and his weight function is shown in Fig. 5.21. It is for the special case of a flowmeter with a uniform field, and it shows how the velocity, at any point in the cross-section of the tube, 'weights' the signal. Thus, far from the electrodes the velocity has about 25 per cent of the effect of flow near the electrodes. This shows that, although an ideal design for an axisymmetric

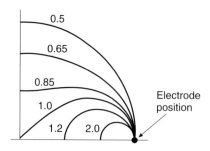

Fig. 5.21 Shercliff weight function (shown for one quadrant)

flow profile, it is far from ideal for a realistic flow profile, say, downstream of a bend.

Magnetic field excitation

The magnetic field coils of this type of flowmeter were, for many years, excited using AC at 50 or 60 Hz. However, in addition to sinusoidal excitation current, a square-wave excitation current, or a combination of these, may now be available and there may be advantages of each which should be discussed with the manufacturer.

A problem with sinusoidal excitation is that a transformer-type spurious voltage is generated and the electrode wires, amplifier, and so on, have to be designed to eliminate this voltage. This voltage and the eddy currents in the pipe walls led in particular to 'zero drift' – the variation of the output at no-flow by 1 per cent or so of the full scale reading. In modern designs this has been largely eliminated and, in addition, many designs have a low-flow cut-off, whereby the signal is set to zero below a certain flowrate.

As a result of these problems, the alternative square-wave excitation was introduced. The actual shape of the wave varies between manufacturers, but essentially it allows sufficient dwell period at two different field excitations to allow spurious voltages to decay and the flow signal to be obtained as a difference between the two levels.

Liner and electrode materials

It is important to obtain advice on material compatibility with the operating liquid both for the liner and the electrodes. Typical liner materials are neoprene, polyurethane, PTFE, and ceramic; typical electrode materials are non-magnetic metals: stainless steel, platinum–iridium, tantalum, hastelloy, etc. Manufacturers may offer an electrode cleaning system, which is an additional feature to retain performance.

Accuracy of electromagnetic flowmeters

Typical values for these meters should be:

- uncertainty after calibration of the order of ±0.5 to ±1.0 per cent of rate or better;
- range turndown ratio from 10:1 possibly up to much greater claimed values.

Installation effects/requirements

Installation guidelines are:

- the error due to a reducer installed next to the flowmeter is likely to be less than 1 per cent;
- an error of up to 2 per cent should be assumed for other fittings separated from the meter (electrode plane) by 5D of straight pipe;
- an error of 1 per cent may occur for other fittings separated from the meter by 10D of straight pipe;
- a pipe fitting at least 3D downstream of the electrode plane should not affect the response;
- the orientation of the flowmeter at 5D spacing or greater does not simply correlate with the size of the error;
- flowmeter orientation should, therefore, be decided on other grounds, such as retaining the electrodes in a horizontal plane to avoid problems with bubbles breaking the circuit.

This probably equates to low to medium sensitivity in Table 3.2, but most of the studies appear to have assumed that a small shift in reading was expected.

The development of the weight function has led to an extensive design method which has allowed progress from the simple flowmeter with a uniform field requiring long axial length to accommodate a long magnet, and resulting in a bulky instrument, to compact flowmeters with low susceptibility to profile effects. Modern designs are unlikely to have a uniform magnetic field, S is unlikely to be unity, and the weight function will give the best compromise to achieve a good performance with the most likely range of disturbed flows. The guidelines above are a simple distillation of the data from modern designs that have been influenced by this design method.

The measurement uncertainty is typical of electromagnetic flowmeters currently available. The effect of upstream pipe fittings is less than for most other types of flowmeter. The flowmeter should be installed with at least 10D upstream between the fitting and the electrode plane to avoid additional uncertainty in excess of 1 per cent. Although theoretical considerations indicate that flowmeter orientation is important, in practice the developing flow from a fitting is so complex that the orientation should be decided on other grounds, notably that the electrodes are kept in a horizontal plane.

Downstream pipework should have negligible effect, and three diameters should be ample.

Where designs are of very short axial length, as in the wafer designs for fitting between flanges, the magnetic and conducting properties of the neighbouring pipework may affect the calibration.

Non-contacting electrode designs offer the possibility of electrode shapes such that the meter is almost insensitive to flow profile. This insensitivity can also be achieved with a square-section pipe bore. Another recent design, referred to as linerless, uses electrodes encapsulated in an insulating material together with field coils to form complete assemblies, two such assemblies being inserted into side standpipes.

Advantages of electromagnetic flowmeters

These are:
- linear response;
- clear bore;
- signal down to zero flow (if required);
- low sensitivity to upstream conditions;
- no moving parts;
- suitable for slurries and multiphase flows.

Disadvantages

- Only suitable for conducting liquids (in current commercial designs). Minimum conductivity typically 5 µS/cm, although special designs with non-contacting electrodes may be available for conductivity down to 0.05 µS/cm. Tap water has a conductivity of about 100 µS/cm.
- Suitable in multiphase flows only where the continuous phase is conducting.

Typical applications

- Water, slurry, conducting chemicals, liquid foods, drinks, sewage, liquid metals, etc.

The advantages need little comment. The signal is usable down to zero flow. Early versions exhibited a drift of a per cent or so at zero flow. However, this appears to be less of a problem with modern designs, and particularly with square-wave excitation. In addition, meters may have a low-flow cut-off. However, few other types of meter are capable of continuing to operate with reliability from full flow to very small flows.

The electromagnetic flowmeter has created a particular niche for itself in slurry flow measurement. Indeed, an early application was for flow measurement in dredging operations. It measures the volume flowrate, assuming that the slurry has the velocity of the conducting phase and fills the whole cross-section.

Although it has been made to operate on non-conducting liquids, it may be available commercially for liquids with conductivity equal to that of pure water and can handle most conducting liquids in industry. Tap water has conductivity of about 100 µS/cm. The actual conductivity is not important, provided it is adequate, uniform, and the electrodes are not 'open-circuited' by air or fouling.

The list of applications is very long. It should be considered for any conductive liquid where flow measurement is required. There can be problems where the tube runs partially empty but some designs have been introduced for this type of application. It is essential that electrodes retain contact with the liquid. Bubbles may interfere with the continuity and hence the reason for installing the meter so that the electrodes are in the horizontal plane. The meter offers an obstruction-free, reliable, electrical-output meter of good accuracy requiring low maintenance.

5.5   ULTRASONIC METERS

There are three primary types of ultrasonic meter which we shall consider.

- The time-of-flight meter which is probably the most accurate of the three, and depends on pulses of ultrasound being transmitted across the pipe at an angle so that they are carried by or retarded by the flow of the fluid. Within this group is the sing-around system.
- The Doppler meter which senses the flow of liquids by Doppler-shifted reflections off particles or other discontinuities in the fluid.
- The cross-correlation meter which times the transit of the liquid by correlation of the signal pattern between two axially spaced sets of transducers.

Some commercial meters may combine more than one of these systems in one meter.

5.5.1   Time-of-flight

Description of operation
The time-of-flight designs are capable of high accuracy. The difference in the time of transit between transducers of upstream and downstream pulses of ultrasound (Fig. 5.22) is used to obtain the flowrate in the tube. However, the signal is dependent on the speed of sound in the fluid. The transit time (or leading edge) system obtains the speed of sound by taking the mean time of transit of the pulses up and downstream.

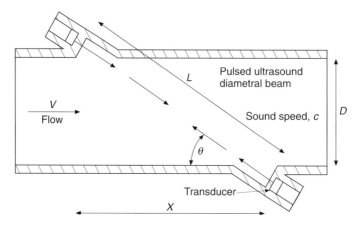

Fig. 5.22 Ultrasonic time-of-flight flowmeter

The sing-around system eliminates the value of the speed of sound, c, by obtaining the sing-around frequencies resulting from pulse trains across the pipe created by the received pulse triggering a transmitted pulse. Thus, as one pulse is received at one transducer it triggers a pulse from the other transducer. The upstream and the downstream pulse trains each have their own frequency, and the difference between the frequencies allows the flowrate to be obtained. In most designs the same pair of transducers operate in both directions.

Relationship between time difference and flowrate – transit time
Two ultrasound pulses are transmitted. The upstream pulse is retarded by the flow and takes a longer time to reach the receiving transducer, given by

$$t_u = \frac{D/\sin\theta}{c - V\cos\theta} \tag{5.7}$$

The downstream pulse is speeded up by the flow and takes a shorter time to reach the receiving transducer, given by

$$t_d = \frac{D/\sin\theta}{c + V\cos\theta} \tag{5.8}$$

From the two values of elapsed time the speed of sound (mean) and the flowrate (difference) are obtained. It is important to obtain the sound speed as the accuracy is directly dependent on it and the sound speed is affected by temperature, fluid parameters, etc.

Putting $\Delta t = t_u - t_d$, cot $\theta = X/D$ and noting that, to a close approximation, $c^2 = L^2/t_u t_d$, the volumetric flowrate is given by

$$q_v = \frac{\pi D^2 L^2 \Delta t}{8 X t_u t_d} \tag{5.9}$$

The accuracy of the instrument will depend on the smallest time which can be measured. To achieve 1 per cent uncertainty, time must be measured for a 100 mm tube to within 1 ns. Typical transit times and differences – where $E(\Delta t)$ is the discrimination in Δt required to achieve an uncertainty of better than 1 per cent and is taken as $\Delta t/100$ – are given in Table 5.1.

Relationship between frequency difference and flowrate – sing-around
For the sing-around system, the upstream frequency is given by

$$f_u = \frac{c - V\cos\theta}{D/\sin\theta} \tag{5.10}$$

and for downstream pulses is given by

$$f_d = \frac{c + V\cos\theta}{D/\sin\theta} \tag{5.11}$$

Putting $\Delta f = f_d - f_u$, the volumetric flowrate is given by

$$q_v = \frac{\pi D^2 L^2 \Delta f}{8X} \tag{5.12}$$

The limitations of the sing-around system relate more to the very low frequencies and long time periods needed to achieve sufficient accuracy. To achieve 1 per cent uncertainty we shall need to have a measuring period of greater than 1 s for the most favourable of these cases. Typical frequencies and differences are given in Table 5.2.

Table 5.1 Typical times and time differences for a transit time ultrasonic flowmeter

Diameter (mm)	$(t_u + t_d)/2$ (s)	Δt (s)	$E(\Delta t)$ (s)
100	10^{-4}	10^{-7}	10^{-9}
300	3×10^{-4}	3×10^{-7}	3×10^{-9}

where $X/D = 1$, $V = 1$ m/s, and c is taken as 1430 m/s, the value for water at 4 °C.

Table 5.2   Typical frequencies and frequency differences for a sing-around ultrasonic flowmeter

Diameter (mm)	V = 1 m/s		V = 10 m/s	
	f_d (Hz)	Δf (Hz)	f_d (Hz)	Δf (Hz)
100	10 117	10	10 162	100
300	3372	3.3	3387	33

where $X/D = 1$, $V = 1$ m/s, and c is taken as 1430 m/s, the value for water at 4 °C.

Flowmeter path/beam positions

For a non-uniform profile the expressions for time and frequency difference are

$$\Delta t = \frac{2 \cot \theta}{c^2} \int V(x) \, dx \tag{5.13}$$

$$\Delta f = \frac{\sin 2\theta}{D^2} \int V(x) \, dx \tag{5.14}$$

For both of these equations, the integration is in the plane of the path of the ultrasound beam and perpendicular to the direction of the pipe axis. This may be a diameter of the pipe, or it may be a chord. This raises a further problem resulting from the fact that the integral across a diameter of a circular pipe is not the integral we need for the mean flow in a circular pipe, but is overweighted by the high velocities at the centre of the pipe. For a single-beam flowmeter with the path across the diameter of the pipe, simple integration of the fully developed turbulent profiles for various Reynolds numbers shows that, for each 10:1 turndown, the diametral integration introduces a non-linearity of about ±0.5 per cent. In addition, single-beam designs are very susceptible to distortions in the flow profile due to upstream fittings. For this reason multiple beams are used to improve the accuracy by integrating across several pipe chords. Two-beam designs have an almost correct integration for fully developed profiles, and they also give a much improved performance in distorted profiles. Instruments with more than two beams are very high accuracy instruments and can achieve high performance based on measurement of path length, angle, etc. without calibration. Figure 5.23 shows some of the paths used with and without reflection of the beam on the inner surface of the pipe.

Transducers

The ultrasound is transmitted and received by piezo-electric crystals (in most designs) which are incorporated into the transducer to achieve certain

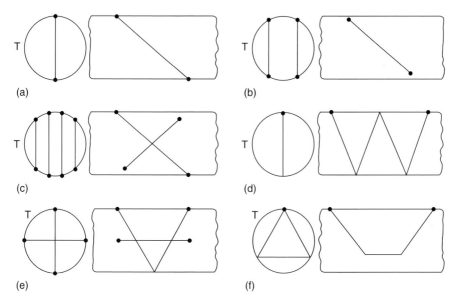

Fig. 5.23 Ultrasonic beam positions (T, top of pipe, indicates possible preferred orientation in liquid flow): (a) diametral path; (b) mid-radius paths; (c) four paths; (d) reflex mode with W path; (e) reflex mode with two V paths; (f) reflex mode with triangular path (from Baker 2000 reproduced by kind permission of Cambridge University Press)

characteristics such as damping, directionality, etc. Various designs of transducer are available including retro-fit (hot tap) and clamp-on transducers which can be used on existing pipework. The latter designs allow the instrument to be installed on the outside of a pipe without breaking into the pipe. It is usually necessary to specify the dimensions and material of the pipe when ordering such an instrument.

Figure 5.24 shows a selection of transducer mountings. For small pipes the fluid is brought into the pipe from the side and flows axially between the transducers. The advantage of this is that a greater length can be used resulting in a greater time difference for up- and downstream pulses than would occur by traversing only a small pipe diameter.

Accuracy of ultrasonic time-of-flight flowmeters

- Probable measurement uncertainty for liquids and gases ±0.5 to 5 per cent of rate.
- For clamp-on flowmeters, uncertainty probably in the range ±3 to 7 per cent of rate.
- Non-linearity of the order of ±1 per cent for 20:1 turndown.
- Range turndown for liquids 10:1 to 20:1 or higher.
- Range turndown for gases 50:1 or higher.

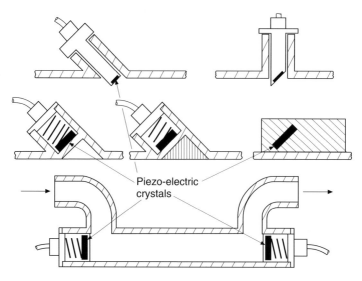

Piezo-electric
crystals

Fig. 5.24 Ultrasonic transducer types

Installation effects/requirements

Installation effects/requirements for high accuracy meters to retain performance to within ±1 per cent of calibration are likely to be as follows:

- One beam – very sensitive to upstream flow distortion and 20 to 40D straight pipe should be allowed upstream of meter;
- Two beam – allow 10 to 20D straight pipe between flowmeter and upstream pipe fittings;
- Four beam – allow 10D upstream.

Referring to Table 3.2, installation effects are likely to be:

- for single beam: medium to high;
- for dual beam: low to medium;
- for more than two beams: low to negligible.

Note: 5D should be allowed downstream to avoid acoustic interference from flow-generated noise.

The value of measurement uncertainty is possibly pessimistic, but there is a wide range of instrument performance. The errors due to installation downstream of a pipe fitting on a single-beam meter can be high, and manufacturers should be consulted on the required upstream straight lengths to retain performance. The two-beam meter is much superior, and a value of 10D may be adequate for most upstream disturbances. The four-beam designs may be even less sensitive to upstream fittings than suggested above.

Advantages of ultrasonic time-of-flight flowmeters

- No obstruction to flow.
- Capability for retrofitting.
- Available in clamp-on designs for liquid and gas.
- Probably unaffected by pulsatile flow.
- Suitable for most single-phase liquids.
- Suitable for gas flows.

Disadvantages of ultrasonic time-of-flight flowmeters

- Transducer cavities may collect air or debris.
- Response time may be slow.
- Unsuitable for most two-phase flows.
- Piezo-electric crystals must be in contact with gas in most designs.

Typical applications

- Water and other liquids.
- Gas flows, including custody transfer of gas.
- Oil flows.

A meter has been offered commercially for measuring the flow of drilling mud, but generally these meters suffer from loss of transmission, if applied to two- or multiphase flows.

5.5.2   Doppler

Description of operation
If acoustic waves reflect off a moving object and return to the source it is found that they experience a frequency shift proportional to the velocity component of the object parallel to the acoustic beam. This shift can be used to obtain the velocity of particles or discontinuities in a flowing fluid. A diagram of such a meter is shown in Fig. 5.25.

Simplest equation
If the transmitted frequency is f_t, then the frequency shift of the reflected signal is

$$\Delta f = 2f_t(V/c) \cos \theta \tag{5.15}$$

This equation incorporates the speed of sound, and this will vary. Some form of compensation should be incorporated in the system. However, the majority of the uncertainty relates to the nature of the reflecting surface. It is not known from which moving object the reflection comes, nor where it is situated in the pipe, nor the speed with which it moves relative to the fluid. Performance may be affected by too few reflectors or by too many. The

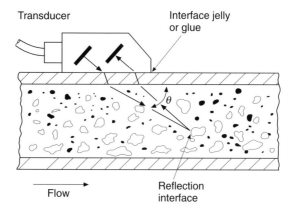

Fig. 5.25 Ultrasonic Doppler flowmeter

uncertainty and the different velocity of fluid and particle or bubble, are most severe in a vertical flow. On occasion, the vibration of the pipe has appeared to be sufficient to provide a (spurious) response. It appears to be preferable for the flow to be multiphase, even if the second phase component is very small, although temperature fluctuations may provide sufficient reflection in some cases. These factors, therefore, all suggest that the device should be viewed as a flow indicator rather than an accurate flowmeter, and in that restricted role, it can provide a useful tool.

Several devices of this type are available and they offer one of the few methods of determining whether or not flow is occurring in a pipe without needing to disturb the pipework. However, they have definite limitations which must be understood. If used intelligently, they can provide the engineer with a useful diagnostic tool.

Accuracy of ultrasonic Doppler flowmeters

- The device is not an accurate one, and its performance is so dependent on installation, that it would be rash to attempt to put a value on the uncertainty.

Installation effects/requirements

- Installation effects are likely to be more in terms of the meter's functioning. Upstream fittings will have an effect which is likely to be lost in the general uncertainty of the instrument.

Advantages

- The meter is easily installed without disturbing the pipework.

Disadvantages

- It requires a second phase even in small quantities and dispersed, or fluid variation.

- Reflecting interfaces, particles, or bubbles may not have the same velocity as the bulk flow, particularly in upward and downward flows.
- Poor mounting combined with vibration may cause spurious readings.

Typical applications

- The meter may be usable as a simple flow monitor where there is no alternative, but sufficient knowledge of the flow to give some confidence in the reading.

In some specialist applications this device has been used with 'range-gating' which provides information as to where the signal reflection has occurred, and may allow integration of the profile.

5.5.3 Cross-correlation

Description of operation
Two diametral acoustic paths (Fig. 5.26a) displaced in the axial direction allow two ultrasound beams to be transmitted and received. The ultrasound, in its passage through the fluid, is disturbed by turbulence or a second phase

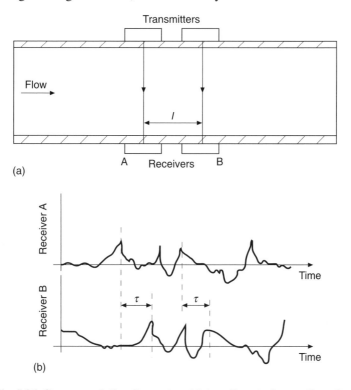

Fig. 5.26 Cross-correlation flowmeter: (a) two diametral acoustic paths; (b) received signals showing time lag in pattern

in the flow. The disturbed signals are correlated, and the time lag in the downstream signal can be obtained as a measure of the transit time of the fluid between the paths (Fig. 5.26b).

Although this description has been in terms of ultrasound, the correlation may use various signals. There are a number of commercial correlators on the market, and these could be applied to a range of phenomena.

Simplest equation

If we take the time lag as τ and the distance between the diametral paths as l, then the velocity is approximated by

$$V = l/\tau \tag{5.16}$$

Since the dominant disturbance causing the variation in the received signals is not known, the relationship with flow profile and phase concentration distribution is not known. However, some attempt has been made to obtain installation effects for such a meter.

Accuracy of ultrasonic cross-correlation flowmeters

- Uncertainties for calibrated instruments are probably between ±2 and 5 per cent of rate, or more if slip occurs between phases/components in the flow.

Installation effects/requirements

Installation effects have been reported for fittings 10D upstream as:

- −4 per cent for single elbow;
- +5 per cent for double elbow out-of-plane;
- −5 per cent for double elbow in-plane.

In terms of Table 3.2 this is likely to rate as high sensitivity.

Advantages

- Possibly suitable for two-phase flows with caution as to what is being sensed.

Disadvantages

- An adequate source of disturbance is necessary so that a correlation signal can be obtained.
- The source of correlation may not be moving with the same velocity as the main fluid, resulting in a further uncertainty in the measured velocity.

Typical applications

- Possibly in two-phase and multiphase flows.

Although this meter appeared promising for two- and multiphase flows, experience has introduced a note of caution. Like the Doppler meter it is not always clear as to the source of correlation. Slip between phases can introduce additional uncertainty. For this reason, use in a vertical flow is inadvisable. It may, therefore, still await a niche application.

CHAPTER 6

Mass Flowmeters

Categorization of mass flow measurement
- Multiple differential pressure meters.
- Mass flow measurement with fluids of known properties.
- Mass flow measurement of multiphase flows.

Thermal mass flow measurement
- Capillary thermal mass flowmeter (CTMF) – gases.
- Capillary thermal mass flowmeter (CTMF) – liquids.
- Insertion and in-line thermal mass flowmeter (ITMF).

Angular momentum fuel mass flowmeter

Coriolis meter

6.1 INTRODUCTION

Categorization of mass flow measurement

A   Direct (true) mass flow measurement – the sensor responds to mass flow.
B   Indirect (inferential) mass flow measurement – separate sensors respond to velocity or momentum and pressure, temperature, etc.

This categorization is somewhat misleading, as single sensors are subject to other effects, while dual sensors may give a true measure of mass flowrate. One meter, which appears at present to approach closely to true mass flow measurement, is the Coriolis meter. Certain meters using heat capacity are sometimes referred to as true mass flowmeters, since heat capacity is mass specific. However, heat capacity is dependent on the fluid and, therefore, the meter will only read correctly if the fluid properties are known. In addition, the heat transfer is viscosity dependent. In contrast, a combination of a volumetric flowmeter and a density meter may be less dependent on fluid properties.

An alternative categorization might be in terms of meters for fluids of 'known' and 'unknown' properties. For the latter, a mass meter would need to be sensitive only to mass flow and unaffected by fluid change, second or third phases, unsteadiness, and so on. It is unlikely that any meter is totally unaffected at present. Therefore, we will review briefly the various types on their merits.

A combination of a volumetric meter of well-documented performance, which is insensitive to fluid properties, with a density meter of the vibrating type, or some other well-tried design, allows mass flow to be deduced. However, the two readings may not refer to the same piece of fluid, which in some applications could be a disadvantage.

6.1.1 Multiple differential pressure meters

Two methods, which will not be discussed here, are the twin venturi method and the method of Brand and Ginsel for which the reader is referred to Baker (2000) for a description and explanation.

Description of operation of the Wheatstone bridge arrangement
The hydraulic Wheatstone bridge method has been used commercially. Figure 6.1 shows two arrangements in the Wheatstone bridge layout. Four matched and identical orifices are arranged in a hydraulic Wheatstone bridge network with a constant volume pump producing a recirculating flow. The process mass flow is q_m and the pump volumetric flow is q_{vp}. The pressure drop is given by Δp_{BD} in Fig. 6.1a for the situation when $q_m > \rho q_{vp}$ (moderate flows). If the flow q_m, were to divide equally between the upper and lower halves of the Wheatstone bridge, then the pressure drop through each half would be

$$\Delta p = K(q_m/2)^2/\rho \tag{6.1}$$

To obtain the equation for the meter system we use

$$\Delta p_{BD} = \Delta p_{AD} - \Delta p_{AB}$$
$$= K(q_m/2 + \rho q_{vp}/2)^2/\rho - K(q_m/2 - \rho q_{vp}/2)^2/\rho$$
$$= K q_m q_{vp}$$

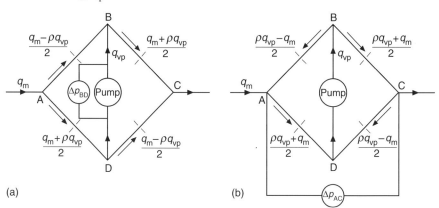

(a)   (b)

Fig. 6.1 Hydraulic Wheatstone bridge flowmeter: (a) high flows; (b) low flows (from Baker 2000 reproduced by kind permission of Cambridge University Press)

So

$$q_m = \frac{\Delta p_{BD}}{K q_{vp}} \qquad (6.2)$$

Similarly for the arrangement in Fig. 6.1b for low flows we obtain

$$q_m = \frac{\Delta p_{AC}}{K q_{vp}} \qquad (6.3)$$

This type of arrangement may well be very sensitive to precise matching of the components, and avoidance of any variation in them.

Accuracy of hydraulic Wheatstone bridge arrangement
- Typical uncertainty after calibration may be of the order of ±0.5 to 1.0 per cent of rate.

Range turndown ratio
- Values claimed up to 100:1 or more.

Installation effects/requirements
- Probably none.

Advantages
- Suitable for fuel flow measurement.

Disadvantages
- Little information available on effects of fluid and environmental changes, maintenance, etc.

Typical applications
- Fuel flow measurement in R&D laboratories, etc.

6.1.2 Mass flow measurement with fluids of known properties

Flowmeters may offer the possibility of conversion from normal use to mass flow by obtaining additional information from existing measurements. There are several examples of this, but we give here one common and one proposed, but not tried, method.

Differential pressure meter
The differential pressure devices require a value for the density, ρ, for both volumetric and mass measurement. Thus the flowmeter equation is

$$q_m = CE\varepsilon \frac{\pi}{4} d^2 \sqrt{(2\rho_1 \Delta p)} \qquad (6.4)$$

where the symbols are defined in Chapter 4, and the required measurements to obtain mass flowrate are the differential pressure, Δp, and the density, ρ_1, which can be obtained either directly through a density measuring cell, or using pressure and temperature to obtain the density for a known gas.

Ultrasonic flowmeter

The transit time ultrasonic flowmeter includes the square of the speed of sound, c^2, in its equation, and we may rearrange the equation to replace the sound speed squared by expressions which include density: $1/(k_s\rho)$ for liquids and $\gamma p/\rho$ for gases.

For a liquid

$$q_m = \frac{\pi D \Delta t}{8k_s \cot\theta} \tag{6.5}$$

and for a gas

$$q_m = \frac{\pi D \gamma p \Delta t}{8 \cot\theta} \tag{6.6}$$

Required measurements are Δt, and for liquid k_s, the adiabatic compressibility, or for gases p, the pressure and γ, the ratio of the specific heats. An alternative approach which is being developed uses the acoustic impedance of the fluid which leads to a value for ρc.

Other possibilities that have been tried

- A turbine meter which measures the drag on the rotor;
- a vortex meter which measures the drag on the bluff body;
- an electromagnetic meter with a gamma-ray densitometer;
- a venturi with a gamma-ray densitometer (possibly at the throat).

Advantages

- Volumetric flowrate known accurately.
- Density may be deduced from existing measurements or measured with a density cell.
- Mass flowrate may be programmed using modern flow computers.

Disadvantages

- Two measurements required, which may not correspond to the same 'piece' of fluid.

6.1.3 Mass flow measurement of multiphase flows

This is too large a subject to be tackled in this introductory guide, but a few points are made and the reader is referred to Baker (2000). The scene is dominated by the oil industry, but other multiphase flows are likely to be significant in the future. There are three main options for the measurement of multiphase flows:

(a) separate the components;
(b) measure main flowrate, but separate a sample to obtain component amounts;
(c) in-line measurement of all parameters.

The oil industry has used option (a) predominantly in the past. As an interim stage in moving to (c), (b) may offer an option. In the longer term, most measurements are likely to move to (c). Examples from the past using option (c) are the nuclear industry for fault flow measurement, while the food industry would, of necessity, use this approach. However:

  (i) the response of metering devices tends to be very sensitive to the local flow regime, which is affected not only by component mass flowrates, but also by the upstream line configuration and flow history. It may therefore, be appropriate to consider upstream flow conditioning;
 (ii) the best practice might be to calibrate the instruments with known phase flowrates and with an exact simulation of upstream pipework but, even so, it may not be possible to cover all phase and flow regimes;
(iii) the response of instruments may be affected by flow transients, including both pressure fluctuations and time-dependent flow regimes;
(iv) if several devices are used together the instruments must be chosen such that the results of any interaction they may have with the flow do not affect the response of a neighbouring instrument.

The most promising approach for measuring multiphase mass flow may be to obtain volumetric flow, and to combine this measurement with that from instruments which obtain phase distribution and density. In addition, line pressure and temperature sensors will probably be required.

For volumetric flowrate of non-conducting fluids some possibilities are:

• flow measurement by venturi, target, cross-correlation, or positive displacement.

For volumetric flowrate of conducting fluids:

• flow measurement by electromagnetic flowmeter.

For density and phase composition:

- acoustic methods;
- electrical impedance methods;
- microwave;
- gamma-ray methods;
- vibrating densitometer etc.

Were any combination of these instruments to be used it would, of course, be necessary to redesign them to be rugged enough for the exacting conditions envisaged. This is particularly true for the important application of sub-sea well-head flow measurement and downhole and production logging.

A flowmeter for these applications may have an uncertainty as great as 5 per cent itself, and it is not, therefore, unreasonable to expect the system uncertainty to be twice this value.

A number of new designs of flowmeter system are being developed for option (c) above.

6.2   THERMAL MASS FLOW MEASUREMENT

The relationship for this family of meters depends on the relationship between heat transfer and temperature rise in a fluid. The heat transfer, Q_h is related to temperature rise in the fluid, specific heat, and mass flowrate by the equation

$$Q_h = K\Delta T c_p q_m \qquad (6.7)$$

where q_m is the mass flowrate, ΔT is the temperature rise, c_p is the specific heat of the fluid, and K is a constant.

6.2.1   Capillary thermal mass flowmeter (CTMF) – gases

Description of operation
Heat is supplied to the gas passing through a capillary tube so that the temperature of the gas rises and the change in temperature between two points provides a measure of the flowrate. In its simplest form the equation for the mass flowrate, q_m, is

$$q_m = \frac{Q_h}{K c_p \Delta T} \qquad (6.8)$$

Figure 6.2 is a diagram of the capillary tube. It shows the heat source passing heat to the tube via a collar on the tube. To each side of this source are thermocouples, TC 1 and TC 2, the temperature difference of which yields ΔT.

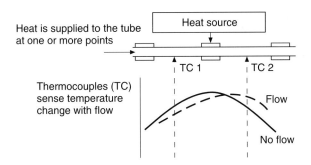

Fig. 6.2 Capillary thermal mass flowmeter (CTMF) – gases

Outside these are two heat sink collars which ensure that the gas temperature is negligibly affected by the presence of the flowmeter. As flow takes place, the temperature distribution in the tube, Fig. 6.2, is distorted by the flow which sweeps the higher temperature region downstream.

The mass flowrate is approximately inversely proportional to the temperature rise. The temperature rise is also dependent on the heat capacity of the gas, which must be known. The specific heat for some gases at constant pressure is given in Table 6.1.

Table 6.1 also indicates that the flowmeter must be calibrated on the correct gas at the operating conditions. Operation in these capillary flowmeters will normally be in laminar flow. The device is likely to be affected by heat transfer rate, and this will be affected, in turn, by the nature of the fluid plus the condition of the pipe. It is sometimes claimed to be a true mass flowmeter, but is too much affected by other parameters for this to be an adequate description.

For larger flows the flowmeter is mounted in a bypass with a laminar flow element in the main pipe to ensure that the flowrate in the bypass is proportional to that in the main pipe. The device is typically of small size (3–6 mm tube) and for low flows unless used with the bypass arrangement.

Table 6.1   Approximate heat capacities of various gases

			Variation with pressure (%)	
Gas	c_p *(kJ/kg K)*	*Variation (% per °C)*	*0–10 bar*	*0–100 bar*
Air	1.01	0.01	2	16
N_2	1.04	0.012	0.1	1.6
O_2	0.92	0.02	1	18
CO_2	0.83	0.004	10	—
CH_4	2.33	0.11	2	31

Accuracy of CTMF – gases

- Typical uncertainty after calibration will, probably, be of order ±1.0 per cent of upper range value (URV).

Range turndown ratio

- Likely to be 15:1 or greater.

Installation effects/requirements for CTMF – gases

- Unlikely to be affected by flow profile, except when used in a bypass arrangement.

Advantages

- Capable of measuring very low flows of gas.

Disadvantages

- Preferably should be calibrated on the correct gas at close to operating conditions.

Typical applications

- Low flows of clean dry gases above their dew point.
- Particularly applied to gas blending, and semi-conductor industries.

6.2.2 Capillary thermal mass flowmeter (CTMF) – liquids

Description of operation

The operation of this device is similar to that of the CTMF for gases, although the geometry will be different to accommodate the heat input and the heat sink arrangements.

Accuracy of CTMF – liquids

- Probable uncertainty after calibration of the order of ±1.0 per cent of URV.

Range turndown ratio

- Of the order of 15:1.

Installation effects/requirements for CTMF – liquids

- Probably no upstream flow effects, but may be sensitive to other environmental parameters.

Advantages

- Low liquid flowrates.

Disadvantages

- Not independent of fluid being measured.

Typical applications of CTMF – liquids

- May be suitable for very low flowrates of toxic, corrosive, and volatile liquids.
- Control and measurement of low flows in laboratories.
- Microfiltration etc.
- Fuel consumption etc.

6.2.3 Insertion and in-line thermal mass flowmeters (ITMF)

While these devices depend on the same physical effects as the CTMF, the design is for an in-line flowmeter, or a probe into a main flow line. For this reason both will be treated together here rather than dealt with in the chapter on probes (Chapter 7). When discussing these devices the word *sensor* will be used for the elements or transducers within the meter which sense temperature and transfer heat.

Description of operation

A typical arrangement is that one sensor measures the temperature of the flowing gas, while the other is retained at about 20° above the temperature of the gas and the heat input to the sensor is measured. The convective heat transfer from the heated sensor is dependent on the flowrate of the gas as well as on the characteristics of the sensor and the gas. If the temperature of the heated sensor changes, the amplifier will adjust the current through the probe to bring its temperature back to the correct setting above the gas temperature. This will determine the heat input and hence the flowrate can be obtained. The relationship for these flowmeters is

$$q_h = k(1 + Kq_m^{0.5})\Delta T \tag{6.9}$$

where q_h is the heat supplied, k is a constant which allows for heat transfer and temperature difference at zero flow, K is a constant incorporating the area of the duct, as well as gas and heat transfer constants, q_m is the mass flowrate and ΔT is the temperature difference between heated and unheated sensors.

The circuitry for such a device is commonly based on a Wheatstone bridge, which is used to adjust the currents and temperatures of the sensors. Some modern designs use microprocessors allowing digital processing for driving the sensors.

Insertion thermal mass flowmeter

The probe is inserted through a suitable fitting in the pipe wall. In Fig. 6.3 a diagram of such a probe is shown with a single pair of sensors, although such a design can have several pairs of sensors spaced across the pipe. Multiple sensors allow some averaging of the flow profile in the pipe.

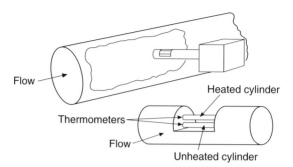

Fig. 6.3 Insertion thermal mass flowmeter (ITMF)

An alternative probe arrangement which has been used for flare gas flow measurement where a wide dynamic range is required, is shown in Fig. 6.4. In this case the flow passes through the probe so that one thermistor is cooled by the flow, while the other senses the temperature of the stagnant gas. Until the introduction of ultrasonic flowmeters for flare gas, this was possibly the most commonly used type of meter in this application.

In-line thermal mass flowmeter
The operation of this device is as for the insertion type, except that, as the name implies, it is permanently mounted in a spool piece. A diagram of such a device is shown in Fig. 6.5. Two sensors are shown in the pipe. Although drawn with one downstream of the other, it is common for them to be in the same diametral plane.

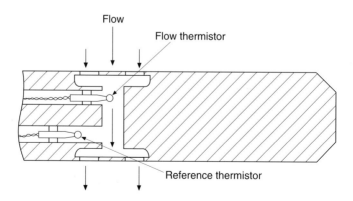

Fig. 6.4 Cross-section of flare gas probe showing typical layout of flow passages and temperature sensors

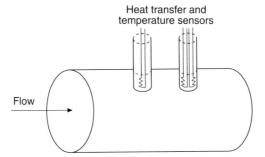

Fig. 6.5 Diagram of an in-line thermal mass flowmeter (ITMF) (the sensors are sometimes mounted in the same cross-section) (from Baker 2000 reproduced by kind permission of Cambridge University Press)

Accuracy for ITMF
- Probable uncertainty after calibration for in-line designs ±1 to 4 per cent URV.
- Probable repeatability for insertion type ±3 per cent of rate.

Range turndown ratio
- Possibly 50:1 or more.

Installation effects/requirements
- Upstream flow effects may be severe, and manufacturers may call for a flow conditioner in the line.
- Referring to Table 3.2, installation effects are likely to be medium to high.

Advantages
- Offers a mass flowmeter for gases (although dependent on gas composition) at a lower price than Coriolis.
- Allows measurement at very low flowrates.

Disadvantages
- Not independent of fluid being measured.

Typical applications
- Process gas flows of many varieties.
- Leakage detection.
- Not usually suitable for steam.

Claims for lower values of uncertainty should be viewed with caution bearing in mind the uncertainties of the calibration facility and the sensitivity of the meter to installation and gas effects. Other things being equal, the insertion meter should have a similar performance, but the cross-sectional area of the flow line will introduce a further uncertainty.

A turndown ratio of 100:1 may be obtainable, but the sensitivity to installation may require caution about the uncertainty over this range.

6.3　Angular momentum fuel flowmeter

Figure 6.6 is a diagram of a commercial design of angular momentum meter. This design consists of two rotors tethered together on a torque shaft. The liquid in the pipeline enters the meter through guide vanes and a drive rotor which extracts a small amount of energy from the flow to drive the rotors. Once through the drive rotor, the flow passes along a section which ensures that the flow is moving axially. The flow emerges from this section axially and enters the main rotor. In doing so the angular momentum of the flowing liquid is changed by the rotor and a torque is applied, as a result, to the rotor. The rotor is restrained on its shaft by a torque spring, and so rotates through an angle due to the torque which has been applied to it. Two pick-ups on the rotor provide a measure of the angle through which the torque-spring part of the rotor has moved compared with the shaft of the rotor. The angular momentum imparted to the liquid by the rotor per second is given approximately by

$$X = \omega R^2 q_{\mathrm{m}} \tag{6.10}$$

where we assume a representative radius of the annulus is R and q_{m} is the mass flowrate. The angular velocity, ω, is known from the impeller speed. Using a long impeller with many vanes and a small annulus this should be a good approximation. The twist created by this should introduce a torque

$$T = s\theta \tag{6.11}$$

where s is the constant of the restraining spring and θ is the deflection angle. Then, since the torque will be equal to the angular momentum imparted to the liquid per second:

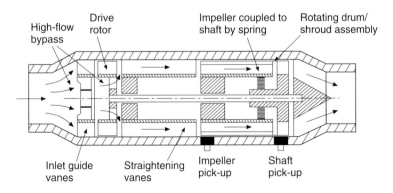

Fig. 6.6　Angular momentum mass flowmeter

$$q_m = \frac{s\theta}{\omega R^2} \qquad (6.12)$$

where s/R^2 will be a constant.

It is important to note the limitations of this device:

(a) the flow will not precisely follow the vanes unless sufficient length is allowed to force the flow to be axial relative to the vanes;
(b) the flow profile in the annulus will modify the effective value of R;
(c) spring stiffness etc. must be stable.

In some designs, the drive turbine allows for changes in the speed of rotation to extend the range of the device. This is achieved by controlling the amount of flow which passes through the driving rotor. In this way a range of nearly 50:1 is achieved.

Accuracy of angular momentum fuel flowmeter
- Probable uncertainty after calibration ± 1.0 per cent of rate for 7:1 turndown.

Range turndown ratio
- Up to nearly 50:1.

Installation effects/requirements
- Upstream flow effects may be low to negligible.

Advantages
- Suitable for aircraft fuel flow measurement.

Disadvantages
- Turbine-type device prone to wear.
- Performance data mainly dependent on manufacturers.

Typical applications
- Aircraft fuel flow transmitter.

6.4   CORIOLIS FLOWMETERS

This type of flow meter has aroused great interest in industry, and a large number of designs have appeared and are continuing to appear to meet this interest. There is always a danger with new types of meter that initial enthusiasm will be replaced by loss of confidence as the device fails to live up to the claims made. There appears to be no indication of this to date. Indeed the instrument is of increasing interest, and with the new generation of straight-tube designs provides a clear bore meter for most applications. It is beginning to meet the need in industry for a mass flowmeter. There are still areas for which it is not suitable, but with the amount of research at present these may well be reduced in the near future.

Figure 6.7 shows some of the various configurations of pipe used to achieve the Coriolis effect on which these meters depend. The fundamental requirement is that a tube rotates while flow takes place. This is achieved, for short periods, during the lateral vibration of the tube (brief periods of rotation due to the flexing of the tube around fixed points at each end of it)

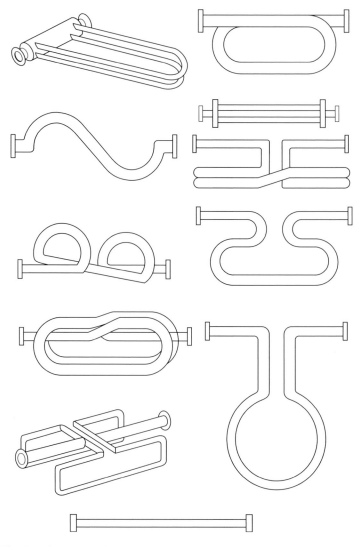

Fig. 6.7 Diagrammatic representations of some configurations of Coriolis mass flowmeters

through which the fluid flows. Although early designs actually used a rotating member, modern designs depend on vibrating the centre of the tube, so that a 'rotation' takes place in each direction as the vibrating tube moves first in one direction and then in the reverse direction. Vibration amplitude can be as small as fractions of a millimetre, and frequencies range from about 75 Hz up to nearly 1000 Hz.

The first commercial Coriolis effect meter was the Micromotion design which achieved this with a U-tube vibrating about a fixed axis so that the flow out along one limb of the U and back along the other, experienced equal and opposite forces twisting the tube.

Figure 6.8 shows two halves of a Coriolis flowmeter tube. These two halves could form a U-tube, or could be attached co-linearly to form a straight-tube meter. Each of the designs in Fig. 6.7 will achieve this 'vibration/rotation' in a similar way.

In Fig. 6.8 the two pieces of pipe are fixed as rigidly as possible at the left-hand ends, and the centre is driven, often with an electromagnetic driver, at a frequency which may be the resonant frequency of the complete tube. As the driver forces the centre of the tube (end of the U-tube or centre of a straight-tube design) upwards, the two halves experience an angular rotation about the fixed ends. When the small mass is at radius r it experiences a lateral velocity, $r\omega$, which increases at $r + \delta r$ to $(r + \delta r)\omega$. The angular momentum of mass, δm, will change from $r^2\,\delta m\omega$ to $(r + \delta r)^2\,\delta m\omega$, a change of $2\delta r\,\delta m\omega$, neglecting δr^2. If δm is moving at velocity V, it takes $\delta r/V$ seconds, and so the inlet half of the pipe will experience a force, $2\omega V\,\delta m$, downwards. The outlet half of the pipe will experience an equal and opposite force. Since the mass depends on the fluid density, $\delta m = \rho A\,\delta r'$, the force will be proportional to $2\omega\rho AV\,\delta r'$. The mass flow, $q_{\mathrm{m}} = \rho AV$, and the force will be related to the mass flowrate by

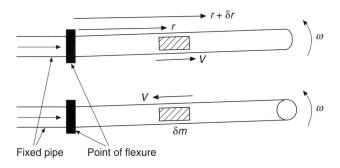

Fixed pipe   Point of flexure

Fig. 6.8 Diagram to illustrate the derivation of equation (6.13)

$$q_\mathrm{m} \propto F/\omega \tag{6.13}$$

The force will distort the tube elastically. In the case of the U-tube, this results in a twist which can be sensed by the time lag in the transit of the two sides of the tube. In the case of a straight tube, sensors at one-quarter and three-quarter positions along the tube will sense the distortion as the tube passes the mid-point.

Thus the effect of vibrating the tube is that the Coriolis forces will cause the inlet half to lag the vibration and the outlet half to lead the vibration. The distortion is in phase with the angular velocity, so it is a maximum as the tube passes the rest position, and is zero at the extremes of deflection of the tube. In most designs two identical pipes mirror each other's vibration, but in the latest generation of straight-tube meters, only one tube is used.

In each case a transducer (magnetic, optical, or similar) is used to sense the distortion or phase difference. A further interesting feature is that the natural frequency of the measuring tube will be related to the *density* of the fluid, and may be used to obtain a value of the density.

The operation of these meters depends on the fixing of the pipe to allow one part to vibrate, while the adjacent pipe is anchored. There is, therefore, a compromise between strength and flexibility. Early problems resulting from metal fatigue, often in the presence of highly corrosive liquids, appear to have been overcome in current designs.

Accuracy of Coriolis meters
- Probable uncertainty after calibration of the following orders of magnitude.
  - For 100 to 50 per cent of flow range: better than ±0.35 per cent of rate.
  - For 50 to 10 per cent of flow range: better than ±0.5 per cent of rate.
  - For 10 to 1 per cent of flow range: better than ±1.0 per cent of rate.

Installation effects/requirements
- Claimed to be insensitive to temperature, pressure, viscosity, and density.
- To a certain extent it may be insensitive to pulsations.
- Negligible effect due to upstream flow distortion.
- Manufacturer's mounting requirements should be followed with care.

Advantages
- Single-tube straight designs provide a clear pipe bore meter to measure mass flow.
- Single-tube straight designs have no more pressure loss than the equivalent straight piece of pipe.
- Industrial designs are likely to include temperature and density measurement.
- No rotating components.
- Available in a range of corrosion-resistant materials.
- Ideal for food and difficult fluid applications.
- Low maintenance.

Disadvantages of Coriolis meters
- Not suitable for two-phase flow although some manufacturers may claim otherwise for certain applications.
- Sensitive to vibration.
- High pressure drop at full flow in some designs due to configuration of internal bends.
- Some designs are bulky.

Typical applications
- Wide range of fluids, both liquid and gas.

The accuracies claimed for these meters are tending to improve, and the values given may prove to be conservative in a short time.

An important requirement for installation is that the user should follow the manufacturer's instructions meticulously to ensure that the meter operates to specification. This may include a vibration-free support structure.

A wide and increasing range of applications are claimed for this meter and it appears to be available in materials compatible with a wide range of fluids.

One less promising application is in two-phase flows, where twin-tube versions split the flow in different phase ratios, and in twin- or single-tube versions the second phase does not necessarily follow the tube vibration correctly. However, for single straight-tube designs the effects may be reduced.

CHAPTER 7

Probes and Tracers

Probes

- Pitot tube.
- Averaging pitot.
- Nozzle.
- Turbine, vortex, electromagnetic, ultrasonic Doppler.
- Ultrasonic transit time.
- Hot wire anemometer.
- Laser Doppler anemometer.

Averaging pitot

Tracers

7.1   PROBES

In Table 1.2 it was shown that, for turbulent pipe flow, a single measurement of velocity could provide a measure of the mean flow in a circular pipe. This is an option which is of interest where the cost of installing a full bore flowmeter cannot be justified. However, there is a conservatism which views the use of a single measurement with some scepticism. This is not entirely misplaced, since there are various possible errors inherent in taking only one measurement. (If the probe is positioned as suggested in Table 1.2, it is in a shear flow, and slight variations in its position or in the turbulence in the flow could affect its reading. If the probe is positioned on the pipe axis, then it will be subject to the change in profile which occurs with changing Reynolds number.) In addition the probe:

(a)   may perform differently in a confined flow;
(b)   may modify the flow in the duct;
(c)   may cause a blockage effect resulting from these and other changes.

   Despite these reservations, there are a range of probes available, as illustrated in Fig. 7.1. There is not space to discuss them in detail, but brief notes are given on each of the devices below.

Pitot tube
This was one of the earliest types, and is still used in laboratory work and in aeronautical applications. The equation is derived in Section 1.7 and is given by equation (1.19). The velocity head in the flow creates a pressure above the

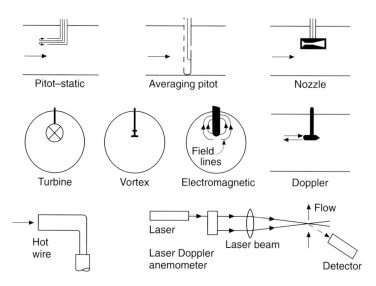

Fig. 7.1 A selection of probes for measuring velocity of flow in a pipe

static pressure, and the difference between pitot and static head allows the flow velocity to be obtained.

Averaging pitot
This device falls between a full bore flowmeter and a probe. Since its main attraction would seem to be as a probe which can be inserted into existing pipework, it is included in this chapter. However, since the main thrust of this book is bulk flow measurement in pipes, it deserves greater mention than the other probes and is covered more fully in Section 7.2.

Nozzle
A further differential device is based on a small nozzle on a probe which operates like one of the devices discussed in Chapter 4.

Turbine, vortex, electromagnetic, ultrasonic Doppler
All these techniques are used for local velocity probes. The turbine is one of the oldest of these devices. However, the electromagnetic probe has become an important option recently with improved designs. Another type of probe is the paddle wheel at the pipe wall which is available commercially.

Ultrasonic transit time
A new device which is inserted into a pipe and then orientated so that it uses transit time signals reflecting off the inner wall of the tube, has recently been proposed and is under development (Rawes and Sanderson 1997).

Hot wire anemometer
This depends on similar principles to the thermal flowmeters discussed earlier. The flow past an electrically heated wire cools it and this can be used to obtain the velocity of the flow in a gas. A similar technique using more robust elements is used in liquids.

Laser Doppler anemometer
The interference between two coherent laser beams which intersect sets up a fringe pattern, and, as particles in the flow cross the fringe pattern, they reflect light with a frequency related to their velocity.

For flow measurement these various techniques will usually be used in a single position probe. However, it is possible to use them to average the flow over the pipe cross-sectional area. In this case special positioning techniques have been developed to obtain the best average of flow in the pipe with the least number of measurement points. This method is used for *in situ* calibration, but may not achieve much better than 5 per cent uncertainty. The use of such a technique requires considerable experience for best results.

7.2   AVERAGING PITOT

Figure 7.2a shows the main features of averaging pitot devices, and typical cross-sections are indicated in Fig. 7.2b. The averaging pitot bar spans the pipe and obtains an approximation to the average total head, based on an average determined by the position of the holes in the tube. Commercially it has various names such as Annubar, Torbar, Kbar, etc. It consists of a shaped bar, usually not round, but rather square or diamond shaped, with about five holes pointing upstream. These holes connect with a plenum space within the bar and the resulting pressure, which is unlikely to be a true average of the five pressures, is carried out to one connection of a differential pressure gauge. At the rear of the bar there are one or more pressure tappings which communicate with a second plenum region in the tube and, in turn, link to the second tapping of the differential pressure gauge. Thus, the difference between the two pressures is related to the flowrate in the pipe. The actual relationship will be provided by the manufacturer of the averaging pitot. Manufacturers of these devices have undertaken extensive tests to identify the effects of installation. However, unless calibrated *in situ*, one major uncertainty in the application of such devices is the cross-section of the pipe into which it is inserted. This will introduce an uncertainty into the calculation of volumetric flowrate, which could be significant.

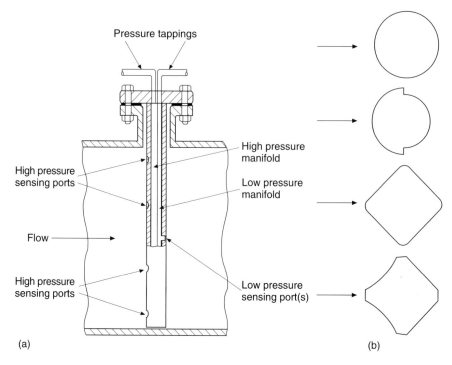

Fig. 7.2 Averaging pitot tube/bar: (a) diagram of a typical meter; (b) cross-section of various designs (from Baker 2000 reproduced by kind permission of Cambridge University Press)

The calibration of the averaging pitot bar will depend on the particular design and the user will be dependent on the manufacturer for information. These devices are also, sometimes, sold already installed in a spool piece, possibly for smaller pipe sizes.

Accuracy of averaging pitot

- Typical uncertainty for the insertion type, is likely to be between ±1 and ±5 per cent allowing for the uncertainty in the cross-section.
- Uncertainty after calibration for the spool piece meter is likely to be ±0.5 to ±1.5 per cent.

Range/turndown ratio

- With smart pressure sensors, 10:1 may be achievable.

Installation effects for averaging pitot

- Upstream installation effects are likely to fall into the highest sensitivity bracket (Table 3.2).
- Manufacturers should have data on sensitivity to upstream conditions, and should be consulted.

Advantages of averaging pitot
- It is possible to insert such a device into existing pipework, and so reduce the cost of the meter and of the installation.

Disadvantages
- Little independent operational information is available on a particular design and it is necessary to rely on the manufacturer's values.

7.3   TRACERS

One of the interesting features of flow measurement is the wide range of techniques which have been used. This is particularly true of tracer methods. They depend either on sensing a disturbance which is naturally in the flow, or on introducing one. Thus, ultrasonic Doppler, correlation techniques, and laser Doppler anemometry, depend on using naturally occurring disturbances or particles in the flow. The tracer methods depend on injection of a chemical, thermal, or nuclear tracer into the flow. This can either be timed or its dilution can be measured.

Table 7.1 gives some of the tracers which have been used, the fluids in which they have been used, and methods of detection. Others have probably been used for specialist applications, and a particular flow may well suggest a particular method due to the particular nature of the fluid.

The transit time method, Fig. 7.3, is an extension of the correlation methods, which, apart from the ultrasound sensing discussed in Section 5.5.3, may use conductivity, capacitance, temperature, optical methods, and probably many other physically measurable parameters.

The concept of introducing a tracer, Fig. 7.3, by upstream injection and then timing it between two downstream stations separated by volume V is an obvious one. If τ is the mean transit time then

$$q_v = V/\tau \qquad\qquad (7.1)$$

Table 7.1   Examples of tracers

Main fluid	Tracer	Detector
Water	Sodium chloride	Electrical conductivity or sample analysis
Liquids	Dyes and chemicals	Fluorimeter, calorimeter, titration
	Temperature change	Temperature sensor
Liquids and gases	Radioactive materials	Nuclear radiation detectors
Gases	SF_6, CO_2, N_2O, He, methane	Infra-red spectrometer

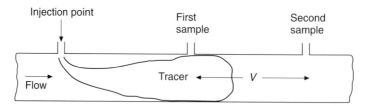

Fig. 7.3 Transit time method using tracer injection upstream followed by measurement at two downstream stations separated by volume *V*

The accuracy of the resultant method is not so obvious, as the distribution of the tracer across the flow may not ensure that the elapsed time is that for the mean movement of the fluid.

The following methods, integration and constant rate, overcome the problem of tracer distribution in pipelines where it is possible to allow time for the tracer to be completely mixed, and where it is possible to measure the concentration of the tracer in the flow. They do make the assumption that the tracer will be uniformly mixed, and there may be applications where this is not so.

In the integration method, Fig. 7.4, fluid of volume v(m^3), with a known concentration of tracer C_1, is injected for a short time and becomes well mixed into the main flow as it passes down the pipe. The fluid is sampled at a suitable position for sufficient time to ensure that all the tracer has passed during the sampling period $\Delta t = t_1 - t_2$. The volumetric flowrate is then obtained from the following expression:

$$q_v = \frac{vC_1}{\int (C_2 - C_0)\,dt} = \frac{vC_1}{\Delta t(\bar{C}_2 - C_0)} \tag{7.2}$$

where C_0 is the concentration in the upstream fluid and C_2 is the concentration at the measuring point. The variation of C_2 with time is shown in Fig. 7.4,

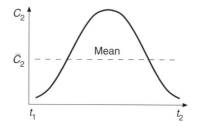

Fig. 7.4 Integration method showing the variation with time of the downstream concentration of the tracer after sudden injection

which also shows the mean value, $\bar{C}_2$. Thus a knowledge of v, Δt, C_0, C_1 and $\bar{C}_2$ allows q_v to be obtained. This method allows small quantities of tracer to be used.

In the constant rate injection method (or dilution method), Fig. 7.5, fluid is added at a constant rate, q_{v1}, with a constant tracer concentration, C_1, to the main flow, q_v, which has a background concentration C_0. Far downstream the final concentration, C_2, is measured. Continuity requires that

$$C_0 q_v + C_1 q_{v1} = C_2 (q_v + q_{v1}) \tag{7.3}$$

so that the volumetric flowrate is

$$q_v = q_1 \frac{C_1 - C_2}{C_2 - C_0} \tag{7.4}$$

Again a knowledge of q_{v1}, C_0, C_1, and C_2 allows q_v to be obtained.

In neither of these methods is it necessary to know the pipe size or the geometry of the pipe, but it is essential that the tracer all flows downstream, and that there are no traps or stagnant regions where tracer might collect. It is also important that the background level of the tracer concentration is a small fraction of the measured value, otherwise the errors in measuring the concentration will become a significant fraction of the concentration difference.

7.4 CONCLUSIONS

The local velocity probes described in this chapter are generally seen as low cost, low accuracy alternatives to a full bore flowmeter.

The averaging pitot may offer a compromise between a full bore flowmeter and sample point velocity measurement. However, its performance and installation sensitivity may only be available from the manufacturer, and knowledge of the exact dimensions of the pipe will also be critical.

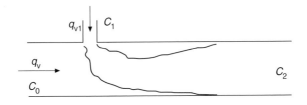

Fig. 7.5 Constant rate tracer injection method (dilution method)

Averaging techniques making use of local velocity probes, and tracer methods may be used for site calibration. However, their use will require special expertise. The total uncertainty achievable with averaging techniques is probably of the order of ±2 to 5 per cent, while for tracer techniques, under good conditions, it may be possible to achieve an uncertainty of the order of ±1 to 2 per cent.

CHAPTER 8

Recent Developments and Likely Future Trends in Flow Measurement

8.1   INTRODUCTION

- Instruments – where are the main developments.
- Meters for multiphase flow.
- Developing technologies.
- Manufacture and management.

If you are a manufacturer with a tight R&D budget, where will you put your money? If you are seeking cost reductions on your manufacture, where will you achieve them? If you are a process plant instrumentation director, what strategy will you have for replacing and upgrading instrumentation? If you are in the R&D business, where are you going to put your efforts? Where can you find answers to these questions?

Over 10 years ago I suggested in the first edition of this *Introductory Guide to Flow Measurement* that the future developments would lie in the areas of:

(a)   non-intrusive, non-invasive, clamp-on, particularly ultrasonic techniques (an area of continuing development);
(b)   optical sensing methods with existing devices (less development than expected);
(c)   intelligent, smart, self-monitoring, and monitoring the plant performance (a dominant area which continues to be of profound importance);
(d)   multiphase flow measurement (continued interest and development, but still some way to go).

An important development, which was evident 10 years ago was the need for new designs and technologies for the utilities. This continues to be an area of development.

In this chapter I shall look at the recent developments and try to see what the future may hold.

8.2 INSTRUMENTS

Orifice plate – improved coefficient of discharge.
Venturi – renewed interest for difficult and multiphase flows.
Critical flow nozzle – much development work going on.

Variable area and other momentum-sensing meters – continued market.

Positive displacement, turbine, vortex – of continuing importance.

Electromagnetic – high reliability and important for the future.
Ultrasonic – increasing accuracy and importance for the future.
Thermal – alternative for gas mass flow measurement.
Coriolis – key instrument for the future.

Probes – A possible cheap alternative.

Orifice plate measurements continue for the discharge coefficient, *C*, and for sources of error, e.g. corrosion, wear, and deposits. The *venturi* is of interest for high pressure gas, wet gas, and other multiphase flows. The *critical flow venturi nozzle* with its potential for very high accuracy is being developed. Other meters such as the *averaging pitot,* the *positive displacement meter,* and the *turbine meter* all retain important roles.

The *vortex meter* appears to have found a particular application in dry-saturated steam. The *electromagnetic flowmeter* has become such a reliable device of high accuracy that its place as a key instrument for the future is assured.

The *ultrasonic meter* is being very actively developed by research centres and companies, and there are reports of increasingly impressive performance.

Several strategies are available for *thermal flowmeters* and development could take place in various directions:

(a) improved theoretical analysis and design of both the capillary thermal mass flowmeter, and the insertion and in-line thermal mass flowmeter designs;
(b) improved theoretical predictions of a meter which is insensitive to flow profile;
(c) use of multiple probes;
(d) use of non-intrusive heating and temperature measurement;
(e) self-checking with transit time to downstream probes.

The *Coriolis flowmeter* straight-tube versions and the meter's use with gases, are likely to dominate the single-phase flowmeter scene for the next few years. The analysis and modelling of its performance and the detailed design is now entering a new phase with the object of ensuring that its claimed performance under wide operational parameters is retained. The manufacturing precision is likely to be increasingly important for this instrument. The signal processing is taking on board the benefits of digital techniques. Multiple sensing may also result from more powerful processing.

Probes offer an alternative in some cases, and recent and new designs of electromagnetic probe and an ultrasonic meter (Rawes and Sanderson 1997) may well set new standards of accuracy.

8.3   METERS FOR MULTIPHASE FLOW

Instruments required for:
- wet gas;
- water-in-oil;
- multiphase flows.

Uncertainty of ±5 to ±10 per cent for components may be achievable for limited ranges.

Interpretation of all variables needed.

Use in other industries as well as oil, e.g. food.

The oil industry demands a meter to deal with the increasingly extreme conditions both of the flow and of the environment. The development of the oil and gas fields and the changing nature and ratios of the components from the wells, have resulted in a need for developing solutions to keep up with these changes. Huge rewards, and substantial expenditure, have ensured that this area has moved forward focused on these moving targets. The instruments required are for:

(a)   wet gas;
(b)   water-in-oil;
(c)   multiphase flow.

The number of multiphase instruments in use may be of the order of 1000 or so, but comparatively few have been on the seabed.

The problems relating to the accurate measurement of multicomponent flow are so great that separation has been the only viable route in the past. Recent developments have exploited multisensor systems combining flowrate with technologies such as single and dual energy gamma-ray densitometers. In 1995 one report suggested, for flows in a vertical line with upward flow, average relative errors of (Hartley *et al.* 1995):

- for liquids 3.9 per cent;
- for gas   7.6 per cent;

- for oil   7.9 per cent;
- for water   5.2 per cent;
- water cut determined to 3.3 per cent.

It would be satisfactory to have a flowmeter that could interpret all the variables occurring in a multicomponent flow. Such a device might identify the position on a flow pattern chart for well-documented fluids. It could also deduce the mass flow of each component. The sophistication needed to achieve this for gases and liquids, without errors due to water, sand, and wax, is high and most methods will need to compromise. One approach may be pattern recognition (neural network) methods where the meter/system 'learns' to interpret signal patterns as flowrates.

The wider uses of multiphase flowmeters should also be remembered, particularly in the food industry.

8.4   DEVELOPING TECHNOLOGIES

Materials.
Sensor technologies.
Micro-devices.
IT in instrumentation, control, and bus protocols.
Use of computational fluid dynamics (CFD) to predict performance and even replace testing.
e-calibration.

Advances in *materials technology* have meant that previously insoluble applications problems, often resulting in an unsatisfactory compromise, are now within the scope of new materials. This applies to new developments in meters such as Coriolis, as well as meter covers etc. for many long-standing designs and instruments.

Sensors depend on vibrating elements, capacitance, resistance, voltage, magnetism, radio frequency, optics, thermal, strain, etc. Modern developments in these and in signal processing call for a review of these methods.

Micro devices (resonating bridge, thermal and Coriolis flow sensors) have appeared seeking applications.

From a mechanical engineer's viewpoint there appear to be a number of fairly obvious ways in which *the power of IT* will continue to influence the instrumentation and control field. They range from the initial interpretation of the signal and its conversion into digital or other form, the transmission and further processing, the linking with a bus system, the interface between the control computers and the people operating the system, and the modelling of the operation of flowing networks.

It has been suggested (Hilgenstock and Ernst 1996) that *CFD solutions* could be as accurate as rig testing, and might be capable of replacing testing in the near future. Areas of particular application could be detailed investigation of flows, installation effects testing, and design prediction for meters such as the critical venturi, averaging pitot, and thermal flowmeter for simple geometries and flows.

The power of *digital methods* to analyse signals and identify modes within them, is being applied to the Coriolis meter. It is also likely to affect increasingly meters such as: electromagnetic, ultrasonic, thermal. The possibilities within multiphase for signal analysis and neural network techniques are likely to result in major developments. The techniques for obtaining flow and wetness information from differential pressure devices with different characteristics are likely to lend themselves to the use of digital analysis of the signals.

Behind such information will be detailed computer graphics. The computer experts, with engineers and scientists, will attempt to model various theories of catastrophic failure, to enable the computer system to recognize possible symptoms at an early stage.

In addition there is *the power of the internet* to keep in contact with customers, and even with instruments sold. There may also be methods developed to allow calibration making use of the internet.

8.5   MANUFACTURE AND MANAGEMENT

Environmental requirements and regulations.
Reducing waste, increasing profits, reducing product failure, attracting larger markets.
Increasing robustness in manufacture and use.
Management of all aspects and stages of flowmeter application and use.

Environmental requirements and regulations will continue to influence flow measurement in areas of the plant that traditionally have been unmonitored, or at best, poorly monitored (Ginesi 1997).

Taking the whole process from identification of product need, through design and manufacture, to marketing, sales, and product maintenance, the next 10 years are likely to see much greater attention paid to all aspects in an attempt to reduce waste, increase profits, reduce product failure, and attract larger markets.

In the area of instrumentation, there is room for careful understanding of the links between production methods and instrument accuracy. It is an area where precise measurement of the production process, and of the product as it moves through the process, is needed, and where new methods are likely to be developed.

The management of flowmeters at all stages, from selection, through application in complex systems, to identifying malfunction, is clearly an area where modern information technology methods would be attractive. How does one select? How do we allow for the costs of ownership? How can we check performance and identify emerging problems? Meter 'fingerprints' have been suggested to deduce the errors in flow measurement which may result in billing problems. However, with the internet the possibility of user feedback and of direct monitoring of the validity of the instrument's performance, would appear to be areas of substantial development.

8.6 CONCLUSIONS

Dominant factors that are increasingly apparent in the industry, have grown in importance over the past 10 years, and are likely to continue to do so in the next 5 years.

- the increasing importance of three modern meters: electromagnetic, ultrasonic, and Coriolis;
- the potential of ultrasonic technology;
- the multiphase systems for oil/gas/water multiphase production, against the changing proportions of oil:gas:water being extracted from wells;
- new materials;
- new sensors for use in flowmeters;
- micro-devices seeking applications;
- the increasing importance of information technology: digital software to interpret sensor signals, powerful algorithms to enhance performance, bus systems to communicate with the meter, computational fluid dynamics as an alternative to experiment;
- the measurement of water and gas for domestic purposes;
- pressures for reduced cost, higher reliability, reduced waste, and improved manufacturing precision.

These points have essentially been covered with the exception of the potential of ultrasonic technology in which this author continues to see untapped potential for the development of a clamp-on meter (possibly a master meter) which might provide an *in situ* calibration, and which senses:

- wall thickness;
- the quality of the inside of the tube;
- the turbulence level;
- profile from a range-gated Doppler system;
- flow measurement from a transit-time system;

- correlation to provide information about a second phase;
- density from the impedance and sound speed;
- condition (self) monitoring.

These areas offer possibilities for new partnerships between the science base and the industry, which should be fruitful in developing a new generation of instrumentation. It is in this area that the Instrumentation Group of the Institute for Manufacturing at Cambridge (http://www.ifm.eng.cam.ac.uk/ig) in collaboration with Cranfield University has sought to provide a resource for the instrumentation industry.

The aim is to provide access to the expertise of Cambridge and Cranfield Universities, with some of the world's foremost university science and technology, and through workshops and evening events to update members on issues of immediate concern to the industry.

Bibliography and References

Some of the books and articles below have been referred to in the text and others provide further reading for those who seek more information than this brief book can contain. I have included in this list books and articles which I have found useful, or which I consider offer the best statement on various topics. In addition to these references, I have found many of the manufacturers' documents of particular value in writing this book and, more generally in my teaching and contract research work. In addition, there is a growing number of software packages available from manufacturers for flowmeter selection. It would be almost impossible to ascribe every piece of information which I have gained from them. In any case, the views are mine, gained over years of teaching and research.

GENERAL

Flow measurement

Baker, R. C. (Ed.). 'Flow'. Special Issue of *Measmt Control*, 1986, **19**(5).
Baker, R. C. Flowmeter. In *Instruments of Science, An Historical Encyclopedia* (Eds R. Bud and D. J. Warner), 1998 (Garland Publishing Inc., New York and London).
Baker, R. C. *Flow Measurement Handbook,* 2000 (Cambridge University Press, New York).
Bean, H. S. (Ed.). *Fluid Meters; their Theory and Application*, 6th edition, 1971 (American Society of Mechanical Engineers, New York).
Endress+Hauser. *Flow Handbook*, English edition, 1989 (Flotec AG, Switzerland).
Furness, R. A. *Fluid Flow Measurement*, 1989 (Longman, Harlow, UK; in association with the Institute of Measurement and Control).
Furness, R. A. and Heritage, J. E. *The Redwood Flowmeter Directory*, 1989 (IBC Technical Services London).
Ginesi, D. Flow sensing: the next generation. *Control Engng*, 1997, November; 56–64.
BS 7405: 1991 *Guide to the Selection and Application of Flowmeters*, 1991 (British Standards Institution, London).
Hayward, A. T. J. *Flowmeters. A Basic Guide and Source Book for Users*, 1979 (Macmillan, Basingstoke).

Herschy, R. W. *Stream Flow Measurement*, 2nd edition, 1995 (E & F N Spon, London).

Hilgenstock, A. and Ernst, R. Analysis of installation effects by means of computational fluid dynamics – CFD vs experiments? *J. Flow Measmt Instrumn,* 1996, **7**(3/4), 161–171.

ISO 4006: 1991 *Measurement of Fluid Flow in Closed Conduits – Vocabulary and Symbols,* 1991 (ISO, Geneva, Switzerland).

Medlock, R. S. The techniques of flow measurement. *Measmt Control,* 1982, **15**, 458–463; 1983, **16**, 9–13.

Miller, R. W. *Flow Measurement Engineering Handbook,* 1996 (McGraw Hill, New York).

National Engineering Laboratory Short Course Notes, The Principles and Practice of Flow Measurement (NEL, East Kilbride).

Ower, E. and Pankhurst, R. C. *The Measurement of Air Flow,* 1966 (Pergamon Press, Oxford).

Scott, R. W. W. (Ed.). *Developments in Flow Measurement,* 1982 (Applied Science, London).

Spitzer, D. W. (Ed.). *Flow Measurement,* 1991 (Instrument Society of America).

Accuracy

BIPM/IEC/IFCC/ISO/IUPAC/IUPAP/OIMI, *Guide to the Expression of Uncertainty in Measurement,* 1st edition, 1993 (ISO, Geneva, Switzerland).

Kinghorn, F. C. The analysis and assessment of data. In *Developments in Flow Measurement* (Ed. R. W. W. Scott), 1982, Ch. 9, p. 141 (Applied Science, London).

Hayward, A. T. J. *Repeatability and Accuracy,* 1977 (Mechanical Engineering Publications, London).

ISO 7066: 1979 *Assessment and Uncertainty in Calibration and Use of Flow Measurement Devices,* 1st edition, 1979 (ISO, Geneva, Switzerland).

ISO 5168: 1978 *Measurement of Fluid Flow – Estimation of Uncertainty in a Flow-rate Measurement,* 1st edition, 1978 (cf. BS 5844) (ISO, Geneva, Switzerland).

UKAS. *The Expression of Uncertainty and Confidence in Measurement,* 1st edition, 1997 (United Kingdom Accreditation Service, Teddington).

Fluid mechanics

Baker, R. C. *An Introductory Guide to Industrial Flow*, 1996 (Professional Engineering Publishing).
Duncan, W. J., Thom, A. S. and Young, A. D. *Mechanics of Fluids* 1970 (Edward Arnold, London).
Miller, D. S. *Internal Flow Systems*, 1978 (BHRA Fluid Engineering, Cranfield).
Tritton, D. J. *Physical Fluid Dynamics*, 2nd edition, 1988 (Oxford University Press).
Ward-Smith, A. J. *Internal Fluid Flow*, 1980 (Clarendon Press, Oxford).

SPECIFIC

Differential pressure

ISO 5167-1: 1997 *Specification for Square-edged Orifice Plates, Nozzles and Venturi Tubes Inserted in Circular Cross-section Conduits Running Full* (cf. BS 1042 Section 1.1), ISO, Geneva, Switzerland.
ISO Amd 1: 1998 (Section 8.3.2.1 Discharge coefficient, C).
Reader-Harris, M. J. and Sattary, J. A. The orifice plate discharge coefficient equation – the equation for ISO 5167-1, 1996. Flow Measurement Memo FL/462, September 1996, Equation 11 (National Engineering Laboratory, East Kilbride, Scotland).
Spink, L. K. *Principles and Practice of Flow Meter Engineering*, 9th edition, 1978 (The Foxboro Company, Foxboro, Massachusetts).

Sonic nozzle

Arnberg, B. T., Britton, C. L. and Seidl, W. F. Discharge coefficient correlations for circular-arc venturi flowmeters at critical (sonic) flow. ASME paper No.
73-WA/FM-8, 1973.
Brain, T. J. S. and Reid, J. Primary calibrations of critical flow venturi nozzles in high-pressure gas. NEL Report No. 666, February, 1980.
ISO 9300: 1990 *Measurement of Gas Flow by Means of Critical Flow Venturi Nozzles*, 1st edition (ISO, Geneva, Switzerland).

Variable area

Baker, R. C. and Sorbie, I. A review of the impact of component variation in the manufacturing process on variable area (VA) flowmeter performance. *J. Measmt Instrumn,* 2001, **12**, 101–112.

Positive displacement

Baker, R. C. and Morris, M. V. Positive displacement meters for liquids. *Trans. Inst. Measmt Control,* 1985, **7** (4), July–September (see Baker 2000 for an updated form of this paper).

Turbine

Baker, R. C. Turbine and related flowmeters: Part I – Industrial practice. *J. Flow. Measmt Instrumn,* 1991, **2**, 147–162.
Baker, R. C. Turbine flowmeters: Part II – Theoretical and experimental published information. *J. Flow. Measmt Instrumn.,* 1993, **4**, 123–144.
Furness, R. A. Turbine flowmeters in *Developments in Flow Measurement – 1* (Ed. R. W. W. Scott), 1982 (Applied Science, London).

Vortex

Zanker, K. J. and Cousins, T. The performance and design of vortex flowmeters. Conference on *Fluid Flow Measurement in the Mid-1970s,* 1977, NEL, East Kilbride, Scotland.

Electromagnetic

Baker, R. C. Electromagnetic flowmeters. In *Developments in Flow Measurement – 1,* (Ed. R. W. W. Scott), 1982 (Applied Science, London).
Hemp, J. and Sanderson, M. L. Electromagnetic flowmeters – a state of the art review, International Conference on *Advances in Flow Measurement Techniques,* Coventry, UK, September 1981.
Shercliff, J. A. *Electromagnetic Flow Measurement,* 1987 (Cambridge University Press).

Ultrasonic

Sanderson, M. L. and Hemp, J. Ultrasonic flowmeters – a review of the state of the art, International Conference on *Advances in Flow Measurement Techniques,* Coventry, UK, September 1981.

Mass

Baker, R. C. Coriolis flowmeters: industrial practice and published information. *J. Flow. Measmt Instrumn,* 1994, **5**, 229–246.
Baker, R. C. and Gimson, C. The effects of manufacturing methods on the precision of insertion and in-line thermal mass flowmeters. *J. Flow Measmt Instrumn,* 2001, **12**, 113–121.
Gast, T. and Furness, R. A. Mass flow measurement technology. NEL Conference, June 1986.
Scanes, P. Mass flowmeters. *Trans. Soc. Inst. Tech.,* 1959, June, 119–123.

Probes

Rawes, W. and Sanderson, M. L. An ultrasonic insertion flowmeter for *in-situ* calibration. In *Ultrasonics in Flow Measurement,* 1997 (Cranfield University, Bedford, England).

Integration methods

ISO 3385 *Measurement of Clean Water Flow in Closed Conduits – Velocity–Area Method Using Current Meters* (ISO, Geneva, Switzerland).

Tracers

ISO 2975 (BS 5857) *Fluid Flow In Closed Conduits, Using Tracers* (ISO, Geneva, Switzerland).
ISO 4053 *Measurement of Gas Flow in Conduits – Tracer Methods* (ISO, Geneva, Switzerland).

Optical instruments

Dickinson, G. Design considerations for optical flowmeter sensors. In 2nd International Conference on *Flow Measurement,* BHRA, London, 1988.

Multiphase flow measurement

Baker, R. C. Measuring multiphase flow. *Chem. Eng,* October 1988, 39–45.
Baker, R. C. Response of bulk flowmeters to multiphase flow. *Proc. Instn Mech. Engrs, Part C, Journal of Mechanical Engineering Science,* 1991, **205**, 217–229.

Hartley, P. E., Roach, G. J., Stewart, D., Watt, J. S., Zastawny, H. W. and Ellis, W. K. Trial of a gamma-ray multiphase flowmeter on the West Kingfish oil platform. *Nucl. Geophys.,* 1995, **9**(6), 533–552.

Other measurements

Hayward, R. W. *Thermodynamic Tables in SI Units,* 1968 (Cambridge University Press).
Kaye, G. W. C. and Laby, T. H. *Tables of Physical and Chemical Constants,* 1973 (Longman, London).
Noltingk, B. E. (Ed.). *Instrumentation Reference Book,* 1988 (Butterworth, London).

Index